# Simula SpringerBriefs on Computing

## Volume 6

Springer and Simula have launched a new book series, *Simula SpringerBriefs on Computing*, which aims to provide introductions to select research in computing. The series presents both a state-of-the-art disciplinary overview and raises essential critical questions in the field. Published by SpringerOpen, all *Simula SpringerBriefs on Computing* are open access, allowing for faster sharing and wider dissemination of knowledge.

Simula Research Laboratory is a leading Norwegian research organization which specializes in computing. The book series will provide introductory volumes on the main topics within Simula's expertise, including communications technology, software engineering and scientific computing.

By publishing the *Simula SpringerBriefs on Computing,* Simula Research Laboratory acts on its mandate of emphasizing research education. Books in this series are published only by invitation from a member of the editorial board.

More information about this series at http://www.springer.com/series/13548

Joakim  Sundnes

# Introduction to Scientific Programming with Python

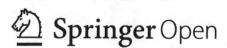

Joakim Sundnes
Simula Research Laboratory
Lysaker, Norway

Simula SpringerBriefs on Computing
ISBN 978-3-030-50355-0        ISBN 978-3-030-50356-7    (eBook)
https://doi.org/10.1007/978-3-030-50356-7

Mathematics Subject Classification (2010): 65D15, 65D25, 65D30, 68-01, 68N01, 68N19, 97-04

This Springer imprint is published by the registered company Springer Nature Switzerland AG
The registered company address is: Gewerbestrasse 11, 6330 Cham, Switzerland

*Dedicated to the memory of Hans Petter Langtangen.*

# Foreword

Dear reader,

Our aim with the series *Simula SpringerBriefs on Computing* is to provide compact introductions to selected fields of computing. Entering a new field of research can be quite demanding for graduate students, postdocs, and experienced researchers alike: the process often involves reading hundreds of papers, and the methods, results and notation styles used often vary considerably, which makes for a time-consuming and potentially frustrating experience. The briefs in this series are meant to ease the process by introducing and explaining important concepts and theories in a relatively narrow field, and by posing critical questions on the fundamentals of that field. A typical brief in this series should be around 100 pages and should be well suited as material for a research seminar in a well-defined and limited area of computing.

We have decided to publish all items in this series under the SpringerOpen framework, as this will allow authors to use the series to publish an initial version of their manuscript that could subsequently evolve into a full-scale book on a broader theme. Since the briefs are freely available online, the authors will not receive any direct income from the sales; however, remuneration is provided for every completed manuscript. Briefs are written on the basis of an invitation from a member of the editorial board. Suggestions for possible topics are most welcome and can be sent to aslak@simula.no.

January 2016

Prof. Aslak Tveito
CEO

Dr. Martin Peters
Executive Editor Mathematics
Springer Heidelberg, Germany

# Preface

This book was originally written as a set of lecture notes to the book *A Primer on Scientific Programming with Python* by Hans Petter Langtangen[1], and can be used either as a supplement to that book or on its own, as a compact introduction to scientific programming. Langtangen's book and these lecture notes, have formed the core of an introductory course on scientific programming at the University of Oslo (INF1100/IN1900, 10 ETCS credits). The course has been running since 2007 and is primarily taken by first-year students of mathematics, engineering, physics, chemistry, and geosciences.

The writing of these lecture notes, and their subsequent evolution into a book, were primarily motivated by two factors. The first was that many students found the nearly 1000 pages of Langtangen's book a bit overwhelming as a first introduction to programming. This effect could be mostly psychological, since the book is well structured and suited for selective study of chapters and sections, but the student feedback from students still indicated the need for a more compact and (literally) lightweight introduction. The second factor was that, sadly, Hans Petter Langtangen passed away in 2016, and his book has therefore not been updated to the newest versions of Python and the various tools introduced in the book. This issue could also be mostly a mental obstacle, since the differences between the Python versions are quite small, and only minor edits are needed to make most of the examples from the original book run on the newest Python platform. However, the book is intended as an introduction to programming, and when learning an entirely new topic, any minor inconsistency is a potential source of confusion. I therefore saw the need for an updated document where all the code examples would run without any modifications on the most common Python platforms. That said, in spite of these minor shortcomings as an introductory text, Langtangen's book is still an excellent resource on scientific programming in Python. Compared with the present book, it covers a much

---

[1]Hans Petter Langtangen, *A Primer on Scientific Programming with Python*, 5th edition, Springer-Verlag, 2016.

broader set of topics and includes more examples, more detailed discussions and explanations, and many more useful programming hints and tips. I highly recommend it as a supplement to these notes for anyone with ambitions to become an expert scientific programmer.

The present book was written specifically for the course *Introduction to programming for scientific applications* (IN1900) at the University of Oslo. It follows exactly the same teaching philosophy and general structure as Langtangen's original book, with the overarching idea that the only way to learn to program is to write programs. Reading theory is useful, but without actual programming practice, the value is very limited. The IN1900 course is therefore largely based on problem solving and programming exercises, and this book's main purpose is to prepare the students for such tasks by providing a brief introduction to fundamental programming concepts and Python tools. The presentation style is compact and pragmatic, and includes a large number of code examples to illustrate how new concepts work and are applied in practice. The examples are a combination of pieces of code (so-called code *snippets*), complete Python programs, and interactive sessions in a Python shell. Readers are encouraged to run and modify the codes to gain a feel for how the various programming concepts work. Source code for most of the examples, as well as Jupyter notebooks for all the chapters, is provided in the online resources accompanying this book.

The typical reader of the book will be a student of mathematics, physics, chemistry, or other natural science, and many of the examples will be familiar to these readers. However, the rapidly increasing relevance of data science means that computations and scientific programming will be of interest to a growing group of users. No typical data science tools are presented in this book, but the reader will learn tasks such as reading data from files, simple text processing, and programming with mathematics and floating point computations. These are all fundamental building blocks of any data science application, and they are essential to know before diving into more advanced and specialized tools.

No prior knowledge of programming is needed to read this book. We start with some very simple examples to get started with programming and then move on to introduce fundamental programming concepts such as loops, functions, if-tests, lists, and classes. These generic concepts are supplemented by more specific and practical tools for scientific programming, primarily plotting and array-based computations. The book's overall purpose is to introduce the reader to programming and, in particular, to demonstrate how programming can be an extremely useful and powerful tool in many branches of the natural sciences.

Many people have contributed to this book, in particular my colleagues at Simula Research Laboratory and the University of Oslo. However, the contributions of Professor Hans Petter Langtangen stand head and shoulders above everyone else. He has been an extremely inspiring teacher, mentor, and colleague throughout my scientific career; he developed the course that is now

IN1900; and he wrote the book on which these notes are based. Throughout these lecture notes I have extensively copied ideas, presentation style, and code examples from his original book, simply because I find them excellent for introducing programming in a scientific context. If it were not for Hans Petter I would clearly never have written these notes. I would probably not be writing this either if he had not, sadly, passed away in 2016 – there would be no need to, because he would surely have written a far better and more extensive book himself.

*May 2020*                                                         *Joakim Sundnes*

# Contents

# Chapter 1
# Getting Started with Python

This book teaches the Python programming language, which is one of the most popular languages for introductory programming courses. An advantage of Python is that it is a so-called high-level language, with simple and intuitive syntax that makes it easy to get started. However, although it works well as a beginner's language, Python is also suitable for more advanced tasks, and it is currently one of the most widely used programming languages worldwide.

## 1.1 The First Example: Hello, World!

Most introductory books on programming start with a so-called Hello, World! program, which is a program that simply writes *Hello, World!* to the screen. In Python, this program is just a single line;

```
print("Hello, World!")
```

To actually write and run such a program, Python offers a number of different options. Throughout this book we will mostly apply the classical programming approach, where a program is written in a text editor and stored as a file that is then run from the command line window or an integrated development environment (IDE). To write and run the "Hello, World!"-program above, open your favorite editor (Atom, gedit, Emacs etc.), type the given line and save the file with a suitable filename, for instance, `hello.py`.[1] Then, open a terminal or an iPython window, navigate to the directory where you saved the file, and type `python hello.py`, if you are using a regular terminal, or `run hello.py` if you are using iPython. The output *Hello, World!* should

---

[1] We do not describe the technical details of acquiring and installing Python here, since this information is platform dependent and becomes outdated very quickly. For updated hints on installing Python, see the web page for the IN1900 course at the University of Oslo (`https://www.uio.no/studier/emner/matnat/ifi/IN1900/index-eng.html`), or to the numerous other resources found online.

© The Author(s) 2020
J. Sundnes, *Introduction to Scientific Programming with Python*, Simula SpringerBriefs on Computing 6,
https://doi.org/10.1007/978-3-030-50356-7_1

appear in the terminal right after the command. If you are using an IDE, it is essentially an editor and an iPython/terminal window combined. For instance, in the popular Spyder IDE the editor is usually in the upper left window, while the window in the lower right corner is the iPython window where you run the program. [2]

Although the "Hello, World!"-program could seem like a silly example, it serves a number of useful purposes. First of all, running this small program will verify that you have installed Python properly, and that you have installed the right version. It also introduces the function `print`, which will be used virtually every time we program, and it illustrates how we use quotes to define a *string* in Python. While `print` is a word that Python understands, the words "Hello" and "World" are not. By using the quotes, we tell Python that it should not try to understand (or interpret) these words, but, rather, treat them as simple text that, in this case, is to be printed to the screen. We will come back to this topic in more detail later.

## 1.2 Different Ways to Use Python

As briefly mentioned above, Python offers some alternatives to the traditional style of programming using a text editor and a terminal window, and some of these alternatives can be very useful when learning to program. For instance, we can use Python interactively by simply typing `python` or `ipython` in a terminal window, without a subsequent file name. This will open an environment for typing and running Python commands, which is not very suitable for writing programs over several lines, but extremely useful for testing Python commands and statements, or simply using Python as a calculator. In a regular terminal window on macOS or Linux, an interactive version of the Hello, World! example would look something like

---

Terminal

```
Terminal> ipython
Python 3.7.3 (default, Mar 27 2019, 16:54:48)
Type 'copyright', 'credits' or 'license' for more information
IPython 7.4.0 -- An enhanced Interactive Python.

In [1]: print("Hello, World!")
Hello, World!

In [2]:
```

---

The two versions `python` and `ipython` work largely the same way, but `ipython` has a number of additional features and is recommended.

---

[2]For details, see, for instance, `https://www.spyder-ide.org/`.

A third way to use Python is through *Jupyter notebooks*, which are a form of interactive notebooks that combine code and text. The notebooks are viewed through a browser and look quite similar to a simple web page, but with the important difference that the code segments are "live" Python code that can be run, changed, and re-run while reading the document. These features are particularly useful for teaching purposes, since detailed explanations of new concepts are easily combined with interactive examples. All the chapters of this book are also available as Jupyter notebooks.

**Minor drawbacks of the Python language.** Although Python is a very popular and suitable language for learning to program, it also has some minor drawbacks. One of the more important is tightly linked to its advantage of being a flexible high-level language with a short and intuitive syntax. Writing small programs in Python can be very efficient, and beginners can quickly start writing useful programs, but the downside is that the code can become messy as the programs grow in size and complexity. Other languages such as C, C++, and Java tend, to enforce more structure in the code, which can be confusing for beginners and annoying when you want to write a small program quickly, but it can be more efficient in the long run when writing larger programs. However, it is certainly possible to write neat and nicely structured programs in Python as well, but this requires a choice by the programmer to follow certain principles of coding style, and is not enforced by the language itself.

Another slightly annoying aspect of Python is that it exists in different versions. At the time of this writing, Python 3 has been dominant for quite a while, but if you look for programming resources online or read older textbooks, you will find many examples using Python 2. For the mathematics-centered programming covered in this book, the difference between Python 2 and Python 3 is actually quite small, but some differences are important to be aware of. The most obvious one is how `print` works. In Python 2, the program above would read `print "Hello, World!"`, that is, without the parentheses. Since nearly all code examples use `print` to some extent, programs written in Python 2 will typically not run in Python 3. One particularly relevant resource for scientific Python (on which this book is largely based) is "A Primer on Scientific Programming with Python", by Hans Petter Langtangen[3]. However, the latest version of that book was written in 2016, and all the code examples are in Python 2 and will stop with an error message if they are run in Python 3. In most cases, the only error is the missing parentheses; so the addition of parentheses to all the print statements will make most of the examples run fine in Python 3. We will comment on some of the other differences between the Python versions later.

---

[3]Hans Petter Langtangen, *A Primer on Scientific Programming with Python*, 5th edition, Springer-Verlag, 2016.

# Chapter 2
# Computing with Formulas

In this chapter, we will go one step beyond the Hello, World! example of the first chapter, and introduce programming with mathematical formulas. Such formulas are essential parts of most programs written for scientific applications, and they are also useful for introducing the concept of *variables*, which is a fundamental part of all programming languages.

## 2.1 Programming Simple Mathematics

To introduce the concepts of this chapter, we first consider a simple formula for calculating the interest on a bank deposit:

$$A = P(1 + (r/100))^n,$$

where $P$ is the initial deposit (the *principal*), $r$ is the yearly interest rate given in percent, $n$ is the number of years, and $A$ is the final amount.

The task is now to write a program that computes $A$ for given values of $P$, $r$ and $n$. We could, of course, easily do so with a calculator, but a small program can be much more flexible and powerful. To evaluate the formula above, we first need to assign values to $P$, $r$ and $n$, and then make the calculation. Choosing, for instance, $P = 100, r = 5.0$, and $n = 7$, a complete Python program that does the calculation and outputs the result reads

```
print(100*(1 + 5.0/100)**7)
```

```
140.71004226562505
```

As described in the previous chapter this line can be typed into an interactive Python session, or written in an editor and stored in a file, for instance `interest0.py`. The program is then run with the command `python`

© The Author(s) 2020
J. Sundnes, *Introduction to Scientific Programming with Python*, Simula SpringerBriefs on Computing 6,
https://doi.org/10.1007/978-3-030-50356-7_2

`interest0.py` in a regular terminal or `run interest0.py` in an iPython window or Spyder.

The `interest0.py` program is not much more complex or useful than the Hello, World! example from the previous chapter, but there are a couple of important differences. First, notice that, in this case we did not use quotation marks inside the parentheses. This is because we want Python to evaluate the mathematical formula, and print the result to the screen, which works fine as long as the text inside the parentheses is valid Python code, or, more precisely, a valid *expression* that can be evaluated to produce a result. If we put quotation marks around the formula above, the code would still work, but the result is not what we want – try it!. At this point, it is also worth noting that, while we stated above that Python is a flexible and high-level language, all programming languages are extremely picky about spelling and grammar. Consider, for instance the line

```
write(100*(1+5,0/100)^7)
```

While most people can read this line quite easily, and interpret it as the same formula as the one above, it makes no sense as a Python program. There are multiple errors: `write` is not a legal Python word in this context, a comma has another meaning than the decimal point, and the hat does not mean exponentiation. We have to be extremely accurate with how we write computer programs, and it takes time and experience to learn this.

The mathematical formula above is evaluated according to the standard rules. The terms are evaluated one by one, from left to right, with exponentiation performed first and then multiplication and division. We use parentheses to control the order of the evaluation, just as we do in regular mathematics. The parentheses around `(1 + 5.0/100)` means that this sum is evaluated first (to obtain 1.05), and then raised to the power of 7. Forgetting the parenthesis and writing `1 + 5.0/100**7` will produce a very different result, just as in mathematics. Because the use of parentheses to group calculations works exactly as in mathematics, it is not very difficult to understand for people with a mathematical background. However, when programming more complicated formulas it is very easy to make mistakes such as forgetting or misplacing a closing parenthesis. This mistake is probably the most common source of error when programming mathematical formulas, and it is worth paying close attention to the order and number of parentheses in the expressions, even for experienced programmers. Getting this principle wrong will lead to either an error message when the code is run or to a program that runs fine but produces unexpected results. The first type of error is usually quite easy to find and fix, but the latter can be much harder.

Although Python is quite strict on spelling and grammar, in programming terms called the *syntax*, there is some flexibility. For instance, whitespace inside a formula does not matter at all. An expression like `5 *2` works just as well as `5*2`. Generally, whitespace in a Python program only matters if it is at the start of a line, which we will return to later. Otherwise, one should use whitespace in order to make the code as readable as possible to humans, since Python will ignore it anyway.

## 2.2 Variables and Variable Types

We are used to variables in mathematics, such as $P$, $r$ and $n$ in the interest formula above. We can use variables in a program too, and this makes the program easier to read and understand:

```
primary = 100
r = 5.0
n = 7
amount = primary * (1+r/100)**n
print(amount)
```

This program spans several lines of text and uses variables, but otherwise performs the same calculations and produces the exact same output as the one-line program above. Still, the use of variables has a few advantages, even in this very simple example. One is that the program becomes easier to read, since the meaning of the numbers becomes more intuitive and the formula is easier to recognize. Another advantage, which could be more important, is that it becomes easier to change the value of one of the variables. This advantage becomes even more obvious in more complex formulas where the same variable occurs multiple times. Having to change the code in multiple places each time a new value is needed is guaranteed to introduce errors. If the same number occurs more than once in a program, it should always be stored in a variable.

The instructions in the program above are called *statements*, and are executed one by one when the program is run. It is common to have one statement per line, although it is possible to put multiple statements on one line, separated by semicolons, as in `primary = 100; r = 5.0; n=7`. For people new to programming, especially those used to reading mathematics, it is worth noting the strict sequence in which the lines are executed. In the mathematical equation above, we first introduced the formula itself, and then defined and explained the variables used in the formula $(P, r, n,$ and $A)$ on the next line. This approach is completely standard in mathematics, but it makes no sense in programming. Programs are executed line by line from the top, so so all the variables must be defined *above* the line where they are used.

The choice of variable names is up to the programmer and, generally, there is great flexibility in choosing such names. In mathematics, it is common to use a single letter for a variable, but a variable in a Python program can be any word containing the letters a–z, A–Z, underscore _ and the digits 0-9, but it cannot start with a digit. Variable names in Python are also case-sensitive, for instance, a is different from A. The following program is identical to the one above, but with different variable names:

```
initial_amount = 100
interest_rate = 5.0
number_of_years = 7
final_amount = initial_amount*(1 + interest_rate/100)**number_of_years
print(final_amount)
```

These variable names are arguably more descriptive, but they also make the formula very long and cumbersome to read. Choosing good variable names is often a balance between being descriptive and conciseness, and the choice can be quite important for making a program easy to read and understand. Writing readable and understandable code is obviously important if you collaborate with others who have to understand your code, but it also makes it easier for you to find errors in the code or develop it further at a later stage. Choosing good variable names is therefore worthwhile, even if you are the only person who will ever read your code.

The program above contains two different types of statements; first there are four *assignment statements*, which assign values to variables, and then a single *print statement* at the end. How these statements work might be quite intuitive, but the assignment statements are worth looking into in more detail. In these statements, the expression on the right-hand side of the equality sign is evaluated first, and then the result is assigned to the variable on the left. An effect of this execution order is that statements such as the following work just fine, and are common in programs:

```
t = 0.6
t = t + 0.1
print(t)
```

```
0.7
```

The line t = t + 0.1 would not make sense as a mathematical equation, but it is a perfectly valid assignment in a computer program. The right-hand side is evaluated first, using the value of t already defined, and then the t variable is updated to hold the result of the calculation. The equality sign in Python is called the *assignment operator*, and, although it works similarly to an equality sign in mathematics, it is not quite the same. If we want the more usual meaning of the equality sign, for instance, to determine if two numbers are equal, the operator to use in Python is ==. A trivial comparison could look like

```
a = 5
print(a == 5)
```

```
True
```

We will see many more such examples later.

**Comments are useful for explaining the thought process in programs.** It is possible to combine the strengths of the two programs above and have both compact variable names and a more detailed description of what each variable means. This can be done using *comments*, as illustrated in the following example:

```
# program for computing the growth of
# money deposited in a bank
primary = 100      # initial amount
r = 5.0            # interest rate in %
n = 7              # the number of years
amount = primary * (1+r/100)**n
print(amount)
```

In this code, all the text following the # symbol is treated as a comment and effectively ignored by Python. Comments are used to explain what the computer instructions mean, what the variables represent, and how the programmer reasoned when writing the program. They can be very useful for increasing readability, but they should not be over-used. Comments that say no more than the code, for instance, a = 5   # set a to 5, are not very useful.

**All variables have types.** So far all the variables we have used have been numbers, which is also how we are used to thinking of variables in mathematics. However, in a computer program we can have many different kinds of variables, not just numbers. More precisely, a variable is a name for a Python *object*, and all objects have a *type*. The type of a variable Python is usually decided automatically based on the value we assign to it. For instance, the statement n = 7 will create a variable of the type *integer*, or int, whereas r = 5.0 will create a variable with type float, representing a floating point number. We can also have text variables, called strings, which have type str. For instance, the Hello, World! example above could have been written as

```
hello = "Hello, World!"
print(hello)
```

Here we create a variable hello, which automatically gets type str, and then print the contents of this variable to the screen. The output is exactly the same as for the first example of Chapter 1.

We can check the type of a variable using the built-in function type:

```
print(type(hello))
print(type(r))
print(type(primary))
print(type(n))
```

```
<class 'str'>
<class 'float'>
<class 'float'>
<class 'int'>
```

We see that the output is as expected from the definitions of these variables above. The word `class` preceding the types indicates that these types are defined as *classes* in Python, a concept we will return to later. It is usually not necessary to check the type of variables inside a Python program, but it could be very useful when learning new concepts or if your program produces errors or unexpected behavior.

We will encounter many more variable types in subsequent chapters. The type of a variable decides how it can be used, and also determines the effects of various operations on that variable. The rules for these operations are usually quite intuitive. For instance, most mathematical operations only work with variable types that actually represent numbers, or they have a different effect on other variable types, when this is natural. For an idea of how this works in Python, think about some simple mathematical operations on text strings. Which of the following operations do you think are allowed, and what are the results: (i) adding two strings together, (ii) multiplying a string with an integer, (iii) multiplying two strings, and (iv) multiplying a string with a decimal number? After giving some thought to this question, check your answers by trying them in Python:

```
hello = "Hello, World!"
print(hello + hello)
print(hello*5)
```

Strings that contain numbers are a potential source of confusion. Consider for instance the code

```
x1 = 2
x2 = "2"
print(x1+x1)
print(x2+x2)
```

```
4
22
```

We see that the variable `x2` is treated as a text string in Python, because it was defined using the quotation marks, even though it contains a single number. For the examples we have seen so far, it is easy to ensure that numbers are numbers, simply by not using quotation marks when they are defined. However, later in this book, we will write programs that read data from files or user input. Such data will usually be in the form of text, and any numbers will be text strings similar to the variable `x2` above. Before using the numbers in calculations, we therefore need to convert them to actual numbers, which can be done with the built-in function `float`:

```
x1 = float(x1)
x2 = float(x2)
print(type(x1))
print(type(x2))
print(x2+x2)
```

```
<class 'float'>
<class 'float'>
4.0
```

Of course, using `float` to convert a string to a number requires that the string actually be a number. Trying to convert a regular word, as in `float(hello)` will make the program stop with an error message. There are numerous other built-in functions for converting between types, such as `int` for conversion to an integer and `str` for conversion to a string. Throughout this book we will mostly use the `float` conversion.

## 2.3 Formatting Text Output

The calculations in the programs above would output a single number, and simply print this number to the screen. In many cases this solution is fine, but sometimes we want several numbers or other types of output from a program. This is easy to do with the `print` function, by simply putting several variables inside the parentheses, separated by comma. For instance, if we want to output both `primary` and `final_amount` from the calculation above, the following line would work:

```
print(primary,final_amount)
```

```
100 140.71004226562505
```

However, although this line works, the output is not very readable or useful. Sometimes a better output format or a combination of text and numbers is more useful, for instance,

```
After 7 years, 100 EUR has grown to xxx EUR.
```

There are multiple ways to obtain this result in Python, but the most recent and arguably most convenient is to use so called *f-strings*, which were introduced in Python 3.6. If you are using an earlier version of Python, the following examples will not work, but there are alternative and fairly similar ways of formatting the text output.

To achieve the output string above, using the *f-string* formatting, we would replace the final line of our program by with

```
print(f"After {n} years, 100 EUR has grown to {amount} EUR.")
```

```
After 7 years, 100 EUR has grown to 140.71004226562505 EUR.
```

There are a couple of things worth noticing here. First, we enclose the output in quotation marks, just as in the Hello, World! example above, which tells Python that this is a string. Second, the string is prefixed with the letter `f`, which indicates that the string is an f-string that could contain something extra. More specifically, the string could contain expressions or variables enclosed in curly brackets, and we have included two such variables, n and amount. When Python encounters the curly brackets inside an f-string, it will evaluate the contents of the curly brackets, which can be an expression or a variable, and insert the resulting value into the string. The process is often referred to as *string interpolation* or *variable interpolation*, and it exists in various forms in many programming languages. In our case, Python will simply insert the current values of the variables n and amount into the string, but, if desired, we can also include a mathematical expression inside the brackets, such as

```
print(f"2+2 = {2+2}")
```

```
2+2 = 4
```

The only requirement for the contents inside the curly brackets is that it be a valid Python expression that can be evaluated to yield some kind of value. Throughout this book we will typically use f-string formatting to insert combining text and numbers, but it may also be used for expressions with other types of output.

The f-string formatting will often produce nicely formatted output by default, but sometimes more detailed control of the formatting is desired. For instance, we might want to control the number of decimal places when outputting numbers. This is conveniently achieved by including a *format specifier* inside the curly brackets. Consider, for instance, the following code:

```
t = 1.234567
print(f"Default output gives t = {t}.")
print(f"We can set the precision: t = {t:.2}.")
print(f"Or control the number of decimals: t = {t:.2f}.")
```

```
Default output gives t = 1.234567.
We can set the precision: t = 1.2.
Or control the number of decimals: t = 1.23.
```

There are many different format specifiers, for controlling the output format of both numbers and other types of variables. We will use only a small subset in this book, and primarily to control the formatting of numbers. In addition to those shown above, the following format specifiers can be useful;

```
print(f"We may set the space used for the output: t = {t:8.2f}.")
```

```
We may set the space used for the output: t =     1.23
```

This specifier is used to control the number of decimals, as well as how much space (the number of characters) used to output the number on the screen. Here we have specified the number to be output with two decimal places and a length of eight, including the decimal places. This form of control is very useful for outputting multiple lines in tabular format, to ensure that the columns in the table are properly aligned. A similar feature can be used for integers:

```
r = 87
print(f"Integer set to occupy exactly 8 chars of space: r = {r:8d}")
```

```
Integer set to occupy exactly 8 chars of space: r =       87
```

Finally, the generic format specifier g outputs a floating point number in the most compact form:

```
a = 786345687.12
b = 1.2345
print(f"Without the format specifier: a = {a}, b = {b}.")
print(f"With the format specifier: a - {a:g}, b = {b:g}.")
```

```
Without the format specifier: a = 786345687.12, b = 1.2345.
With the format specifier: a = 7.86346e+08, b = 1.2345.
```

# 2.4 Importing Modules

We have seen that standard arithmetic operations are directly available in Python, with no extra effort. However, what if more advanced mathematical operations, such as $\sin x$, $\cos x$, $\ln x$, are required? These functions are not available directly, but can be found in a so-called *module*, which must be imported before they can be used in our program. Generally, a great deal of functionality in Python is found in such modules, and we will import one or more modules in nearly all the programs we write. Standard mathematical functions are found in a module named math, and the following code computes the square root of a number using the sqrt function in the math module:

```
import math
r = math.sqrt(2)
# or
from math import sqrt
r = sqrt(2)
# or
from math import *    # import everything in math
r = sqrt(2)
```

This example illustrate three different ways of importing modules. In the first one, we import everything from the math module, but everything we want to use must be prefixed with math. The second option imports only the sqrt function, and this function is imported into the main *namespace* of the program, which means it can be used without a prefix. Finally, the third option imports everything from math into the main namespace, so that all the functions from the module are available in our program without a prefix.

A natural question to ask is why we need three different ways to import a module. Why not use the simple `from math import *` and gain access to all the mathematics functions we need? The reason is that we will often import from several modules in the same program, and some of these modules can contain functions with identical names. In these cases it is useful to have some control over which functions are actually used, either by selecting only what we need from each module, as in `from math import sqrt`, or by importing with `import math` so that all the functions must be prefixed with the module name. To avoid confusion later, it might be good to get into the habit of importing modules in this manner right away, although, in small programs where we import only a single module, there is nothing wrong with `from math import *`.

As another example of computing with functions from `math`, consider evaluating the bell-shaped Gaussian function

$$f(x) = \frac{1}{\sqrt{2\pi}s} \exp\left[-\frac{1}{2}\left(\frac{x-m}{s}\right)^2\right]$$

for $m = 0, s = 2$, and $x = 1$. For this calculation, we need to import the square root, the exponential function, and $\pi$ from the math module, and the Python code may look as follows:

```
from math import sqrt, pi, exp
m = 0
s = 2
x = 1.0
f = 1/(sqrt(2*pi)*s) * exp(-0.5*((x-m)/s)**2)
print(f)
```

Notice that for this more complex formula it is very easy to make mistakes with the parentheses. Such errors will often lead to an error message that points to a syntax error on the *next* line of your program. This can be confusing at first, so it is useful to be aware of. If you obtain an error message pointing to a line directly below a complex mathematical formula, the source is usually a missing closing parenthesis in the formula itself.

**Finding information about Python modules.** At this point, it is natural to ask how we know where to find the functions we want. Say we need to compute with complex numbers. How can we know if there is a module in Python for this? And, if so, what is it called? Generally, learning about the useful modules and what they contain are part of learning Python programming, but knowing where to find such information could be even more important. An excellent source is the Python Library Reference (https://docs.python.org/3/library/), which contains information about all the standard modules that are distributed with Python. More generally, a Google search for `complex numbers python` quickly leads us to the `cmath` module, which contains mostly the same functions as `math`, but with support for complex numbers. If we know the name of a module and want to check its contents, we can go to straight to the Python Library Reference, but there are also other options. The command `pydoc` in the terminal window can be used to list information about a module (try, e.g., `pydoc math`), or we can import the module in a Python program and list its contents with the built-in function `dir`.

```
import math
print(dir(math))
```

```
['__doc__', '__file__', '__loader__', '__name__', (...) ]
```

## 2.5 Pitfalls When Programming Mathematics

Usually, the mathematical operations described above work as expected. When the results are not as expected, the cause is usually a trivial error introduced during typing, typically assigning the wrong value to a variable or mismatching the number of parentheses. However, some potential error sources are less obvious and are worth knowing about, even if they are relatively rare.

**Round-off errors give inexact results.** Computers have inexact arithmetic because of rounding errors. This is usually not a problem in computations, but in some cases it can cause unexpected results. Let us, for instance, compute $1/49 \cdot 49$ and $1/51 \cdot 51$:

```
v1 = 1/49.0*49
v2 = 1/51.0*51
print(f"{v1:.16f} {v2:.16f}")
```

The output with 16 decimal places becomes

```
0.9999999999999999 1.0000000000000000
```

Most real numbers are represented inexactly on a computer, typically with an accuracy of 17 digits. Neither $1/49$ nor $1/51$ are represented exactly, and the error is approximately $10^{-16}$. Errors of this order usually do not matter, but there are two particular cases in which they can be significant. In one case, errors can accumulate through numerous computations, ending up as a significant error in the final result. In the other case, which is more likely to be encountered in the examples of this book, the comparison of two decimal numbers can be unpredictable. The two numbers `v1` and `v2` above are both supposed to be equal to one, but look at the result of this code:

```
print(v1 == 1)
print(v2 == 1)
```

```
False
True
```

We see that the evaluation works as expected in one case, but not the other, and this is a general problem when comparing floating point numbers. In most cases the evaluation works, but in some cases it does not. It is difficult or impossible to predict when it will not work, and the behavior of the program thus becomes unpredictable. The solution is to always compare floats by using a tolerance value, as in

```
tol = 1e-14
print(abs(v1-1) < tol)
print(abs(v2-1) < tol)
```

```
True
True
```

There is no strict rule for setting the value of the tolerance `tol`; however, it should be small enough to be considered insignificant for the application at hand, but larger than the typical machine precision $10^{-16}$.

**Some words are reserved and cannot be used as variables.** Although the choice of variable names is up to the programmer, some names are reserved in Python and are not allowed to be used. These names are `and`, `as`, `assert`, `break`, `class`, `continue`, `def`, `del`, `elif`, `else`, `except`, `exec`, `finally`, `for`, `from`, `global`, `if`, `import`, `in`, `is`, `lambda`, `not`, `or`, `pass`, `print`, `raise`, `return`, `try`, `with`, `while`, and `yield`. Memorizing this list is by no means necessary at this point, and we will use many of these reserved words in our programs later, so it will become quite natural to not use them as variable names. However, for programming physics and mathematics, it could be worth noting `lambda`, since the Greek letter $\lambda$ is common in physics and mathematics formulas. Since Python does not understand Greek letters, it is common to just spell them out when programming a formula, that is, $\alpha$ becomes `alpha`, and so on. However, using this approach for $\lambda$ will lead to an error, and the error message might not be very easy to understand. The problem is easily solved by introducing a small intentional typo and writing `lmbda` or similar.

**Integer division can cause surprising errors.** In Python 2, and many other programming languages, unintended *integer division* can sometimes cause surprising results. In Python 3 this is no longer a problem, so you are not likely to run into it during this course, but it is worth being aware of, since many other programming languages behave in this way. Recall from above that various operations behave differently, depending on the type of the variable they work on, such as in adding two strings versus adding numbers. In Python 2, the division operator, `/`, behaves as in normal division if one of the two arguments is a float, but, if both are integers then it will perform integer division and discard the decimal portion of the result. Consider the following interactive session, which runs Python 2.7:

```
————————————————————————  Terminal  ————————————————
Terminal> python2.7
Python 2.7.14 (default, Sep 22 2017, 00:06:07)
(...)
>>> print(5.0/100)    #the parentheses are optional in Python 2.7
0.05
>>> print(5/100)
0
```

Integer division is useful for many tasks in computer science, and is therefore the default behavior of many programming languages, but it is usually not what we want when programming mathematical formulas. Therefore, it could be a good habit to ensure that variables used in calculations are actually floats, by simply defining them as $r$ = 5.0 rather than $r$ = 5. Although it does not really make a difference in Python 3, it is good to get into this habit simply to avoid problems when programming in other languages later.

# Chapter 3
# Loops and Lists

In this chapter, programming starts to become useful. The concepts introduced in the previous chapter are essential building blocks in all computer
programs, but our example programs only performed a few calculations,
which we could easily do with a regular calculator. In this chapter, we will
introduce the concept of *loops*, which can be used to automate repetitive and
tedious operations. Loops are used in most computer programs, and they look
very similar across a wide range of programming languages. We will primarily
use loops for calculations, but as you gain more experience, you will be able
to automate other repetitive tasks. Two types of loops will be introduced in
this chapter: the `while` loop and the `for` loop. Both will be used extensively
in all subsequent chapters. In addition to the loop concept, we will introduce
Boolean expressions, which are expressions with a true/false value, and a new
variable type called a list, which is used to store sequences of data.

## 3.1 Loops for Automating Repetitive Tasks

To start with a motivating example, consider again the simple interest calculation formula;

$$A = P \cdot (1 + (r/100))^n.$$

In Chapter 2 we implemented this formula as a single-line Python program,
but what if we want to generate a table showing how the invested amount
grows with the years? For instance, we could write $n$ and $A$ in two columns
like this

```
0    100
1    105
2    110
3    ...
...  ...
```

© The Author(s) 2020
J. Sundnes, *Introduction to Scientific Programming with
Python*, Simula SpringerBriefs on Computing 6,
https://doi.org/10.1007/978-3-030-50356-7_3

How can we make a program that writes such a table? We know from the previous chapter how to generate one line in the table:

```
P = 100
r = 5.0
n = 7
A = P * (1+r/100)**n
print(n,A)
```

We could then simply repeat these statements to write the complete program:

```
P =100; r = 5.0;
n=0;   A = P * (1+r/100)**n;   print(n,A)
n=1;   A = P * (1+r/100)**n;   print(n,A)
...
n=9;   A = P * (1+r/100)**n;   print(n,A)
n=10;  A = P * (1+r/100)**n;   print(n,A)
```

This is obviously not a very good solution, since it is very boring to write and errors are easily introduced in the code. As a general rule, when programming becomes repetitive and boring, there is usually a better way of solving the problem at hand. In this case, we will utilize one of the main strengths of computers: their strong ability to perform large numbers of simple and repetitive tasks. For this purpose, we use *loops*.

The most general loop in Python is called a while loop. A while loop will repeatedly execute a set of statements as long as a given condition is satisfied. The syntax of the while loop looks like the following:

```
while condition:
    <statement 1>
    <statement 2>
    ...
<first statement after loop>
```

The `condition` here is a Python expression that is evaluated as either true or false, which, in computer science terms, is called a Boolean expression. Notice also the indentation of all the statements that belong inside the loop. Indentation is the way Python groups code together in blocks. In a loop such as this one, all the lines we want to be repeated inside the loop must be indented, with exactly the same indentation. The loop ends when an unindented statement is encountered.

To make things a bit more concrete, let us use write a while loop to produce the investment growth table above. More precisely, the task we want to solve is the following: Given a range of years $n$ from zero to 10, in steps of one year, calculate the corresponding amount and print both values to the screen. To write the correct while loop for solving a given task, we need to answer four key questions: (i) Where/how does the loop start, that is, what are the initial values of the variables; (ii) which statements should be repeated inside the loop; (iii) when does the loop stop, that is, what condition should become false to make the loop stop; and (iv) how should variables be updated for each pass of the loop? Looking at the task definition above, we should be able to answer all of these questions: (i) The loop should start at zero years, so our initial condition should be n = 0; (ii) the statements to be repeated are the evaluation of the formula and the printing of n and A; (iii) we want the loop to stop when n reaches 10 years, so our condition becomes something like n <= 10; and (iv) we want to print the values for steps of one year, so we need to increase n by one for every pass of the loop. Inserting these details into the general while loop framework above yields the following code:

```
P = 100
r = 5.0
n = 0
while n <= 10:              # loop heading with condition
    A = P * (1+r/100)**n    # 1st statement inside loop
    print(n, A)             # 2nd statement inside loop
    n = n + 1               # last statement inside loop
```

The flow of this program is as follows:

1. First, n is 0, $0 \leq 10$ is true; therefore we enter the loop and execute the loop statements:

   - Compute A
   - Print n and A
   - Update n to 1

2. When we have reached the last line inside the loop, we return to the while line and evaluate $n \leq 10$ again. This condition is still true, and the loop statements are therefore executed again. A new A is computed and printed, and n is updated to the value of two.

3. We continue this way until n is updated from 10 to 11; now, when we return to evaluate $11 \leq 10$, the condition is false. The program then jumps straight to the first line after the loop, and the loop is finished.

*Useful tip:* A very common mistake in while loops is to forget to update the variables inside the loop, in this case forgetting the line n = n + 1. This error will lead to an infinite loop, which will keep printing the same line forever. If you run the program from the terminal window it can be stopped with Ctrl+C, so you can correct the mistake and re-run the program.

## 3.2 Boolean Expressions

An expression with a value of true or false is called a Boolean expression. Boolean expressions are essential in while loops and other important programming constructs, and they exist in most modern programming languages. We have seen a few examples already, including comparisons such as a == 5 in

Chapter 2 and the condition `n <= 10` in the while loop above. Other examples of (mathematical) Boolean expressions are $t = 140$, $t \neq 140$, $t \geq 40$, $t > 40$, $t < 40$. In Python code, these are written as

```
t == 40   # note the double ==, t = 40 is an assignment!
t != 40
t >= 40
t >  40
t <  40
```

Notice the use of the double `==` when checking for equality. As we mentioned in Chapter 2 the single equality sign has a different meaning in Python (and many other programming languages) than we are used to from mathematics, since it is used to assign a value to a variable. Checking two variables for equality is a different operation, and to distinguish it from assignment, we use `==`. We can output the value of Boolean expressions with statements such as `print(C<40)` or in an interactive Python shell, as follows:

```
>>> C = 41
>>> C != 40
True
>>> C < 40
False
>>> C == 41
True
```

Most of the Boolean expressions we will use in this course are of the simple kind above, consisting of a single comparison that should be familiar from mathematics. However, we can combine multiple conditions using `and/or` to construct while loops such as these:

```
while condition1 and condition2:
    ...

while condition1 or condition2:
    ...
```

The rules for evaluating such compound expressions are as expected: `C1 and C2` is `True` if both `C1` and `C2` are `True`, while `C1 or C2` is `True` if at least one of the two conditions `C1` and `C2` is `True`. One can also negate a Boolean expression using the term `not`, which simply yields that `not C` is `True` if `C` is `False`, and vice versa. To gain a feel for compound Boolean expressions, you can go through the following examples by hand and predict the outcome, and then try to run the code to obtain the result:

```
x = 0;  y = 1.2
print(x >= 0 and y < 1)
print(x >= 0 or y < 1)
print(x > 0 or y > 1)
print(x > 0 or not y > 1)
print(-1 < x <= 0)    # same as -1 < x and x <= 0
print(not (x > 0 or y > 0))
```

Boolean expressions are important for controlling the flow of programs, both in while loops and in other constructs that we will introduce in Chapter 4. Their evaluation and use should be fairly familiar from mathematics, but it is always a good idea to explore fundamental concepts such as this by typing in a few examples in an interactive Python shell.

## 3.3 Using Lists to Store Sequences of Data

So far, we have used one variable to refer to one number (or string). Some-times we naturally have a collection of numbers, such as the $n$-values (years) $0, 1, 2, \ldots, 10$ created in the example above. In some cases, such as the one above, we are simply interested in writing all the values to the screen, in which case using a single variable that is updated and printed for each pass of the loop works fine. However, sometimes we want to store a sequence of such variables, for instance, to process them further elsewhere in the program. We could, of course, use a separate variable for each value of n, as follows:

```
n0 = 0
n1 - 1
n2 = 2
...
n10 = 10
```

However, this is another example of programming that becomes extremely repetitive and boring, and there is obviously a better solution. In Python, the most flexible way to store such a sequence of variables is to use a list:

```
n = [0, 1, 2, 3, 4, 5, 6, 7, 8, 9, 10]
```

Notice the square brackets and the commas separating the values, which is how we tell Python that n is a list variable. Now we have a single variable that can hold all the values we want. Python lists are not reserved just for numbers and can hold any kind of object, and even different kinds of objects. They also have a great deal of convenient built-in functionality, which makes them very flexible and useful and extremely popular in Python programs.

We will not cover all the aspects of lists and list operations in this book, but we will use some of the more basic ones. We have already seen how to initialize a list using square brackets and comma-separated values, such as

```
L1 = [-91, 'a string', 7.2, 0]
```

To retrieve individual elements from the list, we can use an index, for instance L1[3] will pick out the element with index 3, that is, the fourth element (having a value of zero) in the list, since the numbering starts at zero. List indices start at zero and run to the $n-1$, where $n$ is the number of elements in the list:

```
mylist = [4, 6, -3.5]
print(mylist[0])
print(mylist[1])
print(mylist[2])
len(mylist)  # length of list
```

The last line uses the built-in Python function `len`, which returns the number of elements in the list. This function works on lists and any other object that has a natural length (e.g., strings), and is very useful.

Other built-in list operations allow us, for instance, to append an element to a list, add two lists together, check if a list contains a given element, and delete an element from a list:

```
n = [0, 1, 2, 3, 4, 5, 6, 7, 8]
n.append(9)    # add new element 9 at the end
print(n)
n = n + [10, 11]      # extend n at the end
print(n)
print(9 in n)         #is the value 9 found in n? True/False
del n[0]              #remove the first item from the list
```

These list operations, in particular those to initialize, append to, and index a list, are extremely common in Python programs, and will be used throughout this book. It is a good idea to spend some time making sure you fully understand how they work.

It is also worth noting one important difference between lists and the simpler variable types we introduced in Chapter 2. For instance, two statements, such as a = 2; b = a would create two integer variables, both having value 2, but they are not the same variable. The second statement b=a will create a copy of a and assign it to b, and if we later change b, a will not be affected. With lists, the situation is different, as illustrated by the following example:

```
>>> l[0] = 2
>>> a = [1,2,3,4]
>>> b = a
>>> b[-1] = 6
>>> a
[1, 2, 3, 6]
```

Here, both a and b are lists, and when b changes a also changes. This happens because assigning a list to a new variable does not copy the original list, but instead creates a *reference* to the same list. So a and b are, in this case, just two variables pointing to the exact same list. If we actually want to create a copy of the original list, we need to state this explicitly with b = a.copy().

# 3.4 Iterating Over a List with a for Loop

Having introduced lists, we are ready to look at the second type of loop we will use in this book: the for loop. The for loop is less general than the while loop, but it is also a bit simpler to use. The for loop simply iterates over elements in a list, and performs operations on each one:

```
for element in list:
    <statement 1>
    <statement 2>
    ...
<first statement after loop>
```

The key line is the first one, which will simply run through the list, element by element. For each pass of the loop, the single element is stored in the variable `element`, and the block of code inside the for loop typically involves calculations using this `element` variable. When the code lines in this block are completed, the loop moves on to the next element in the list, and continues in this manner until there are no more elements in the list. It is easy to see why this loop is simpler than the while loop, since no condition is needed to stop the loop and there is no need to update a variable inside the loop. The for loop will simply iterate over all the elements in a predefined list, and stop when there are no more elements. On the other hand, the for loop is slightly less flexible, since the list needs to predefined. The for loop is the best choice in most cases in which we know in advance how many times we want to perform a set of operations. In cases in which this number is not known, the while loop is usually the best choice.

For a concrete for loop example, we return to the investment growth example introduced above. To write a for loop for a given task, two key questions must be answered: (i) What should the list contain, and (ii) what operations should be performed on the elements in the list? In the present case, the natural answers are (i) the list should be a range of $n$-values from zero to 10, in steps of 1, and (ii) the operations to be repeated are the computation of A and the printing of the two values, essentially the same as in the while loop. The full program using a for loop thus becomes

```
years = [0, 1, 2, 3, 4, 5, 6, 7, 8, 9, 10]
r = 5.0
P = 100.0
for n in years:
    A = P * (1+r/100)**n
    print(n, A)
```

As with the while loop, the statements inside the loop must be indented. Simply by counting the lines of code in the two programs shows that the for loop is somewhat simpler and quicker to write than the while loop. Most people will argue that the overall structure of the program is also simpler and less error-prone, with no need to check a criterion to stop the loop or to update any variables inside it. The for loop will simply iterate over a given list, perform the operations we want on each element, and then stop when it reaches the end of the list. Tasks of this kind are very common, and for loops are extensively used in Python programs.

The observant reader might notice that the way we defined the list `years` in the code above is not very scalable to long lists, and quickly becomes repetitive and boring. As stated above, when programming become repetitive and boring, a better solution usually exists. Such is the case here, and very rarely do values in a list need to be filled explicitly, as done here. Better alternatives include a built-in Python function called `range`, often in combination with a for loop or a so-called *list comprehension*. We will return to these tools later in the chapter. When running the code, one can also observe that the two columns of degrees values are not perfectly aligned, since `print` always uses the minimum amount of space to output the numbers. If we want the output in two nicely aligned columns, this is easily achieved by using the f-string formatting we introduced in the previous chapter. The resulting code can look like this:

```
years = [0, 1, 2, 3, 4, 5, 6, 7, 8, 9, 10]
for n in years:
    r = 5.0
    P = 100.0
    A = P * (1+r/100)**n
    print(f'{n:5d}{A:8.2f}')
```

Output is now nicely aligned:

```
    0  100.00
    1  105.00
    2  ...
```

**A for loop can always be translated to a while loop.** As described above, a while loop is more flexible than a for loop. A for loop can always be transformed into a while loop, but not all while loops can be expressed as for loops. A for loop always traverses traverses a list, carries out some processing on each element, and stops when it reaches the last one. This behavior is easy to mimic in a while loop using list indexing and the `len` function, which were both introduced above. A for loop of the form

```
for element in somelist:
    # process element
```

translates to the following while loop:

```
index = 0
while index < len(somelist):
    element = somelist[index]
    # process element
    index += 1
```

**Using the function range to loop over indices.** Sometimes we do not have a list, but want to repeat the same operation a given number of times. If we know the number of repetitions this task is an obvious candidate for a for loop, but for loops in Python always iterate over an existing list (or a list-like object). The solution is to use a built-in Python function named range, which returns a list of integers[1]:

```
P = 100
r = 5.0
N = 10
for n in range(N+1):
    A = P * (1+r/100)**n
    print(n,A)
```

Here we used range with a single argument $N+1$ which will generate a list of integers from zero to $N$ (not including $N+1$). We can also use range with two or three arguments. The most general case range(start, stop, inc) generates a list of integers start, start+inc, start+2*inc, and so on up to, *but not including*, stop. When used with just a single argument, as above, this argument is treated as the stop value, and range(stop) is short for range(0, stop, 1). With two arguments, the interpretation is range(start,stop), short for range(start,stop,1). This behavior, where a single function can be used with different numbers of arguments, is common both in Python and many other programming languages, and makes the use of such functions very flexible and efficient. If we want the most common behavior, we need only provide a single argument and the others are automatically set to default values; however, if we want something different, we can do so easily by including more arguments. We will use the range function in combination with for loops extensively through this book, and it is a good idea to spend some time becoming familiar with it. A good way to gain a feel for how the range-function works is to test statements such as print(list(range(start,stop,inc))) in an interactive Python shell, for different argument values.

**Filling a list with values using a for loop.** One motivation for introducing lists is to conveniently store a sequence of numbers as a single variable, for instance, for processing later in the program. However, in the code above, we did not really utilize this, since all we did was print the numbers to the screen, and the only list we created was a simple sequence from zero to 10. It could be more useful to store the amounts in a list, which can be easily be achieved with a for loop. The following code illustrates a very common way to fill lists with values in Python:

---

[1]In Python 3, range does not technically produce a list, but a list-like object called an iterator. In terms of use in a for loop, which is the most common use of range, there is no practical difference between a list and an iterator. However, if we try, for instance, print(range(3)) the output does not look like a list. To obtain output that looks like a list, which can be useful for debugging, the iterator must be converted to an actual list: print(list(range(3))).

```
P = 100
r = 5.0
N = 10
amounts = []                    # start with empty list
for n in range(N+1):
    A = P*(1+r/100)**n
    amounts.append(A)           # add new element to amounts list
print(amounts)
```

The parts worth noting in this code are `amounts = []`, which simply creates
a list with no elements, and the use of the `append` function inside the for loop
to add elements to the list. This simple way of creating a list and filling it
with values is very common in Python programs.

**Mathematical sums are implemented as for loops.** A very common
example of a repetitive task in mathematics is the computation of a sum, for
instance,

$$S = \sum_{i=1}^{N} i^2.$$

For large values of $N$ such sums are tedious to calculate by hand, but they
are very easy to program using `range` and a for loop:

```
N = 14
S = 0
for i in range(1, N+1):
    S += i**2
```

Notice the structure of this code, which is quite similar to the way we filled
a list with values in the previous example. First, we initialize the summation
variable (S) to zero, and then the terms of the sum are added one by one
for each iteration of the for loop. The example shown here illustrates the
standard recipe for implementing mathematical sums, which are common in
scientific programming and appear frequently in this book. It is worthwhile
spending some time to fully understand and remember how such sums is
implemented.

**How can we change the elements in a list?** In some cases we want to
change elements in a list. Consider first a simple example where we have a list
of numbers, and want to add the value of two to all the numbers. Following
the ideas introduced above, a natural approach is to use a for loop to traverse
the list, as follows:

```
v = [-1, 1, 10]
for e in v:
    e = e + 2
print(v)
```

```
[-1, 1, 10]   # unaltered!!
```

As demonstrated by this small program, the result is not what we want. We added the value of two to every element, but after the loop finished, our list v was unchanged. The reason for this behavior is that although the list is traversed as desired when we create the for loop using for e in v:, the variable e is an ordinary (int) variable, and it is in fact a *copy* of each element in the list, and not the actual element. Therefore, when we change e, we change only the copy and not the actual list element. The copy is overwritten in the next pass of the loop anyway, so, in this case, all the numbers that are incremented by two are simply lost. The solution is to access the actual elements by indexing into the list:

```
v = [-1, 1, 10]
for i in range(len(v)):
    v[i] = v[i] + 2
print(v)
```

```
[1, 3, 12]
```

Notice in particular the use of range(len(v)), which is a common construction in Python programs. It creates a set of integers running from zero to len(v)-1 that can be iterated over with the for loop and used to loop through all the elements in the list v.

**List comprehensions for compact creation of lists.** Above, we introduced one common way of constructing lists, which is to start with an empty list and use a for loop to fill it with values. We can extend this example to fill several lists in one loop, for instance, if we want to examine the effect of low and high interest rates on our bank deposit. We start with two empty lists and fill both with values in the same loop:

```
P = 100
r_low = 2.5
r_high = 5.0
N = 10
A_high = []
A_low = []
for n in range(N+1):
    A_low.append(P*(1+r_low/100)**n)
    A_high.append(P*(1+r_high/100)**n)
```

This approach to using a for loop to fill a list with values is so common in Python that a compact construct has been introduced, called a *list comprehension*. The code in the previous example can be replaced by the following:

```
P = 100
r_low = 2.5
r_high = 5.0
N = 10
A_low = [P*(1+r_low/100)**n for n in range(N+1)]
A_high = [P*(1+r_high/100)**n for n in range(N+1)]
```

The resulting lists `A_low` and `A_high` are exactly the same as those from the for loop, but the code is obviously much more compact. To an experienced Python programmer, the use of list comprehensions also makes the code more readable, since it becomes obvious that the code creates a list, and the contents of the list are usually easy to understand from the code inside the brackets. The general form of a list comprehension looks like

```
newlist = [expression for element in somelist]
```

where `expression` typically involves `element`. The list comprehension works exactly like a for loop; it runs through all the elements in `somelist`, stores a copy of each element in the variable `element`, evaluates `expression`, and appends the result to the list `newlist`. The resulting list `newlist` will have the same length as `somelist`, and its elements are given by `expression`. List comprehensions are important to know about, since you will see them frequently when reading Python code written by others. They are convenient to use for the programming tasks covered in this book, but not strictly necessary, since the same thing can always be accomplished with a regular for loop.

**Traversing multiple lists simultaneously with zip.** Sometimes we want to loop over two lists at the same time. For instance, consider printing out the contents of the `A_low` and `A_high` lists of the example above. We can accomplish this using `range` and list indexing, as in

```
for i in range(len(A_low)):
    print(A_low[i], A_high[i])
```

However, a built-in Python function named `zip` provides an alternative solution, which many consider more elegant and "Pythonic":

```
for low, high in zip(A_low, A_high):
    print(low, high)
```

The output is exactly the same, but the use of `zip` makes the for loop more similar to the way we traverse a single list. We run through both lists, extract the elements from each one into the variables `low` and `high`, and use these variables inside the loop, as we are used to. We can also use `zip` with three lists:

```
>>> l1 = [3, 6, 1];  l2 = [1.5, 1, 0];  l3 = [9.1, 3, 2]
>>> for e1, e2, e3 in zip(l1, l2, l3):
...     print(e1, e2, e3)
...
3 1.5 9.1
6 1 3
1 0 2
```

Lists traversed with `zip` typically have the same length, but the function also works for lists of different lengths. In this case, the for loop will simply stop when it reaches the end of the shortest list, and the remaining elements of the longer lists are not visited.

# 3.5 Nested Lists and List Slicing

As described above, lists in Python are quite general and can store *any* object, including another list. The resulting list of lists is often referred to as a *nested list*. Instead of storing the amounts resulting from the low and high interest rates above as two separate lists, we could put them together in a new list:

```
A_low = [P*(1+2.5/100)**n for n in range(11)]
A_high = [P*(1+5.0/100)**n for n in range(11)]

amounts = [A_low, A_high]   # list of two lists

print(amounts[0])      # the A_low list
print(amounts[1])      # the A_high list
print(amounts[1][2])   # the 3rd element in A_high
```

The indexing of nested lists illustrated here is quite logical, but can take some time getting used to. The important thing is that, if `amounts` is a list containing lists, then, for instance, `amounts[0]` is also a list and can be indexed in the way we are used to. Indexing into this list is done in the usual way, such that, for instance, `amounts[0][0]` is the first element of the first list contained in `amounts`. Playing a bit with indexing nested lists in the interactive Python shell is a useful exercise to understand how they are used.

Iterating over nested lists also works as expected. Consider, for instance, the following code

```
for sublist1 in somelist:
    for sublist2 in sublist1:
        for value in sublist2:
            # work with value
```

Here, `somelist` is a three-dimensional nested list, that is, its elements are lists, which, in turn, contain lists. The resulting nested for loop looks a bit complicated, but it follows exactly the same logic as the simpler for loops used above. When the outer loop starts, the first element from `somelist` is copied into the variable `sublist1`, and then we then enter the code block inside the loop, which is a new for loop that will start traversing `sublist1`, that is, first copying the first element into the variable `sublist2`. Then the process is repeated, with the innermost loop traversing all the elements of `sublist2`, copying each element into the variable `value`, and doing some calculations with this variable. When it reaches the end of `sublist2`, the innermost for loop is over, we "move outward" one level in terms of the loops, to the loop `for sublist2 in sublist`, which moves to the next element and starts a new run through the innermost loop.

Similar iterations over nested loops can be obtained by looping over the list indices, as follows:

```
for i1 in range(len(somelist)):
    for i2 in range(len(somelist[i1])):
```

```
    for i3 in range(len(somelist[i1][i2])):
        value = somelist[i1][i2][i3]
        # work with value
```

Although their logic is the same as regular (one-dimensional) for loops, nested loops look more complicated and it can take some time to fully understand how they work. As noted above, a good way to obtain such understanding is to create some examples of small nested lists in a Python shell or a small Python program, and examine the results of indexing and looping over the lists. The following code is one such example. Try to step through this program by hand and predict the output before running the code and checking the result:

```
L = [[9, 7], [-1, 5, 6]]
for row in L:
    for column in row:
        print(column)
```

**List slicing is used to extract parts of a list.** We have seen how we can index a list to extract a single element, but sometimes it is useful to capture parts of a list, for instance, all the elements from an index $n$ to an index $m$. Python offers *list slicing* for such tasks. For a list A, we have seen that a single element is extracted with A[n], where n is an integer, but we can also use the more general syntax A[start:stop:step] to extract a *slice* of A. The arguments resemble those of the range function, and such a list slicing will extract all elements starting from index start up to but not including stop, with a step step. As for the range function, we can omit some of the arguments and rely on default values. The following examples illustrate the use of slicing:

```
>>> a = [2, 3.5, 8, 10]
>>> a[2:]    # from index 2 to end of list
[8, 10]

>>> a[1:3]   # from index 1 up to, but not incl., index 3
[3.5, 8]

>>> a[:3]    # from start up to, but not incl., index 3
[2, 3.5, 8]

>>> a[1:-1]  # from index 1 to next last element
[3.5, 8]

>>> a[:]     # the whole list
[2, 3.5, 8, 10]
```

Note that these sublists (slices) are *copies* of the original list. A statement such as, for instance, b = a[:] will make a copy of the entire list a, and any subsequent changes to b will not change a. As for the nested lists considered above, a good way to become familiar with list slicing is to create a small list in the interactive Python shell and explore the effect of various slicing operations. It is, of course ,possible to combine list slicing with nested lists, and the results can be confusing even to experienced Python programmers. Fortunately, we will consider only fairly simple cases of list slicing in this book, and we will work mostly with lists of one or two dimensions (i.e., non-nested lists or the simplest lists-of-lists).

## 3.6 Tuples

Lists are a flexible and user-friendly way to store sequences of numbers, and are used in nearly all Python programs. However, a few other data types are also made to store sequences of data. One of the most important ones is called a *tuple*, and it is essentially a constant list that cannot be changed. A tuple is defined in almost the same way as a list, but with normal parentheses instead of the square brackets. Alternatively, we can skip the parentheses and just use a comma-separated sequence of values to define a tuple. The following are two examples that are entirely equivalent and define the same tuple:

```
>>> t = (2, 4, 6, 'temp.pdf')      # define a tuple
>>> t =  2, 4, 6, 'temp.pdf'       # can skip parentheses
```

Tuples also provide much of the same functionality as lists, including indexing and and slicing:

```
>>> t = t + (-1.0, -2.0)           # add two tuples
>>> t
(2, 4, 6, 'temp.pdf', -1.0, -2.0)
>>> t[1]                           # indexing
4
>>> t[2:]                          # subtuple/slice
(6, 'temp.pdf', -1.0, -2.0)
>>> 6 in t                         # membership
True
```

However, tuples are *immutable*, which means that they cannot be changed. Therefore, some operations we are used to from lists will not work. Continuing the interactive session from above, the following are some examples of illegal tuple operations:

```
>>> t[1] = -1
...
TypeError: 'tuple' object does not support item assignment

>>> t.append(0)
...
AttributeError: 'tuple' object has no attribute 'append'

>>> del t[1]
```

```
...
TypeError: 'tuple' object doesn't support item deletion
```

The observant reader might wonder why the line t = t + (-1.0, -2.0) in
the example above works, since t is supposed to be immutable and therefore
impossible to change. The answer is related to the way assignment statements
work in programming. As briefly explained in Chapter 2, assignment works by
first evaluating the expression on the right hand side, which in this example
means to add two tuples together. The result is a new tuple, and neither t
nor (-1.0, 2.0) are changed in the process. Then, the new tuple is *assigned*
to the variable t, meaning that the original tuple is replaced by the new
and longer tuple. The tuple itself is never changed, but the contents of the
variable t is replaced with a new one.

A natural question to ask, then, is why do we need tuples at all, when lists
can do the same job and are much more flexible? The main reason for this is
that, in many cases, it is convenient to work on item that is constant, since it
is protected against accidental changes and can be used as a key in so-called
*dictionaries*, an important Python datastructure that will be introduced in
Chapter 7. Throughout this book, we will not do much explicit programming
with tuples, but we will run into them as part of the modules we import and
use, so it is important to know what they are.

# Chapter 4
# Functions and Branching

This chapter introduces two fundamental programming concepts: *functions* and *branching*. We are used to *functions* from mathematics, where we typically define a function $f(x)$ as some mathematical expression of $x$, and then we can then evaluate the function for different values of $x$, plot the curve $y = f(x)$, solve equations of the kind $f(x) = 0$, and so on. A similar function concept exists in programming, where a function is a piece of code that takes one or more variables as input, carries out some operations using these variables, and produces output in return. The function concept in programming is more general than in mathematics, and is not restricted to numbers or mathematical expressions, but the general idea is exactly the same.

*Branching*, or if-tests, is another fundamental concept that exists in all common programming languages. The idea is that decisions are made in the code based on the value of some Boolean expression or variable. If the expression evaluates to true, one set of operations is performed, and if the expression is false, a different set of operations is. Such tests are essential for controlling the flow of a computer program.

## 4.1 Programming with Functions

We have already used a number of Python functions in the previous chapters. The mathematical functions from the `math` module are essentially the same as we are used to from mathematics or from pushing buttons on a calculator:

```
from math import *
y = sin(x)*log(x)
```

Additionally, we used a few non-mathematical functions, such as `len` and `range`

```
n = len(somelist)
for i in range(5, n, 2):
```

J. Sundnes, *Introduction to Scientific Programming with Python*, Simula SpringerBriefs on Computing 6,
https://doi.org/10.1007/978-3-030-50356-7_4

(...)

and we also used functions that were bound to specific objects, and accessed with the dot syntax, for instance, `append` to add elements to a list:

```
C = [5, 10, 40, 45]
C.append(50)
```

This last type of function is quite special, since it is bound to an object, and operates directly on that object (`C.append` changes `C`). These bound functions are also referred to as *methods*, and will be considered in more detail in Chapter 8. In the present chapter we will primarily consider regular, unbound, functions. In Python, such functions provide easy access to already existing program code written by others (e.g., `sin(x)`). There is plenty of such code in Python, and nearly all programs involve importing one or more modules and using pre-defined functions from them. One advantage of functions is that we can use them without knowing anything about how they are implemented. All we need to know is what goes in and what comes out, and the function can thus be used as a black box.

Functions also provide a way of reusing code we have written ourselves, either in previous projects or as part of the current code, and this is the main focus of this chapter. Functions let us delegate responsibilities and split a program into smaller tasks, which is essential for solving all problems of some complexity. As we shall see later in this chapter, splitting a program into smaller functions is also convenient for testing and verifying that a program works as it should. We can write small pieces of code that test individual functions and ensure that they work correctly before putting the functions together into a complete program. If such tests are done properly, we can have some confidence that our main program works as expected. We will return to this topic towards the end of the chapter.

So how do we write a function in Python? Starting with a simple example, consider the previously considered mathematical function

$$A(n) = P(1 + r/100)^n.$$

For given values $P = 100$ and $r = 5.0$, we can implement this in Python as follows:

```
def amount(n):
    P = 100
    r = 5.0
    return P*(1+r/100)**n
```

These two lines of code are very similar to the examples from Chapter 3, but they contain a few new concepts that are worth noting. Starting with the first line, `def amount(n):` is called the function *header*, and defines the function's interface. All function definitions in Python start with the word `def`, which is simply how we tell Python that the following code defines a function. After `def` comes the name of the function, followed by parentheses containing the function's *arguments* (sometimes called parameters). This simple function takes a single argument, but we can define functions that take multiple arguments by separating the arguments with commas. The parentheses need to be there, even if we do not want the function to take any arguments, in which case we would just leave the parentheses empty.

The lines following the function header are the *function body*, which need to be indented. The indentation serves the same purpose as for the loops in Chapter 3: to specify which lines of code belong inside the function, or to the function body. The two first lines of the function body are regular assignments, but since they occur inside a function, they define *local variables* P and r. Local variables the argument n are used inside the function just as regular variables. We will return to this topic in more detail later. The last line of the function body starts with the keyword `return`, which is also new in this chapter and is used to specify the output returned by the function. It is important not to confuse this return statement with the print statements we used previously. The use of `print` will simply output something to the screen, while `return` makes the function provide an output, which can be thought of as a variable being passed back to the code that called the function. Consider for instance the example `n = len(somelist)` used in the previous chapter, where `len` returned an integer that was assigned to a variable n.

Another important thing to note about the code above is that it does not do much. In fact, a function definition does essentially nothing before it is *called*.[1] The analogue to the function definition in mathematics is to simply write down a function $f(x)$ as a mathematical expression. This defines the function, but there is no output until we start evaluating the function for some specific values of $x$. In programming, we say that we *call* the function when we use it. When programming with functions, it is common to refer to the *main program* as basically every line of code that is not inside a function. When running the program, only the statements in the main program are executed. Code inside function definitions is not run until we include a call to the function in the main program. We have already called pre-defined functions like `sin, len`, etc, in previous chapters, and a function we have written ourselves is called in exactly the same way:

```
def amount(n):
    P = 100
    r = 5.0
    return P*(1+r/100)**n

year1 = 10
a1 = amount(year1)                          # call
a2 = amount(5)                              # call
```

---

[1] This is not entirely true, since defining the function creates a function object, which we can see by defining a dummy function in the Python shell and then calling `dir()` to obtain a list of defined variables. However, no visible output is produced until we actually call the function, and forgetting to call the function is a common mistake when starting to program with functions.

```
print(a1, a2)
print(amount(6))                        # call
a_list = [amount(year) for year in range(11)] #multiple calls
```

The call `amount(n)` for some argument n returns a `float` object, which essentially means that `amount(n)` is replaced by this `float` object. We can therefore make the call `amount(n)` everywhere a `float` can be used.

Note that, unlike many other programming languages, Python does not require the type of function arguments to be specified. Judging from the function header only, the argument of `amount(n)` above could be any kind of variable. However, by looking at how n is used inside the function, we can tell that it must be a number (integer or float). If we write complex functions where the argument types are not obvious, we can insert a comment immediately after the header, a so-called *doc string*, to tell users what the arguments should be. We will return to the topic of doc strings later in this chapter.

## 4.2 Function Arguments and Local Variables

Just as in mathematics, we can define Python functions with more than one argument. The formula above involves both $P$ and $r$ in addition to $n$, and including them all as arguments could be useful. The function definition could then look like

```
def amount(P, r, n):
    return P*(1+r/100.0)**n

# sample calls:
a1 = amount(100, 5.0, 10)
a2 = amount(10, r= 3.0, n=6)
a3 = amount(r= 4, n = 2, P=100)
```

Note that we are using the arguments P, r, and n inside the function exactly as in the previous example, where we defined P and r inside the function. Inside a function, there is no distinction between such *local variables* and the arguments passed to the function. The arguments also become local variables, and are used in exactly the same way as any variable we define inside the function. However, there is an important distinction between *local* and *global* variables. Variables defined in the main program become global variables, whereas variables defined inside functions are local. The local variables are only defined and available inside a function, whereas global variables can be used everywhere in a program. If we tried to access P, r, or n (e.g., by `print(P)`) from outside the function, we will simply obtain an error message stating that the variable is not defined.

**Arguments can be *positional arguments* or *keyword arguments*.** Notice also the alternative ways of calling a function. We can either specify the argument names in the call, as in r=3.0, n=6, or simply pass the values. If we specify the names, the order of the arguments becomes arbitrary, as in the last call above. Arguments that are passed without specifying the name are called *positional arguments*, because their position in the argument list determines the variable to which they are assigned. Arguments that are passed including the name are called *keyword arguments*. Keyword arguments need to match the definition of the function; that is, calling the function above with amount(100, 5.0, year=5) would cause an error message because year is not defined as an argument to the function. Another rule worth noting is that a positional argument cannot follow a keyword argument; a call such as amount(100, 5.0, n=5) is fine, but amount(P=100, 5.0, 5) is not and the program will stop with an error message. This rule is quite logical, since a random mix of positional and keyword arguments would make the call very confusing.

**The difference between local and global variables.** The distinction between local and global variables is generally important in programming, and can be confusing at first. As stated above, the arguments passed to a function, as well as variables we define inside the function, become local variables. These variables behave exactly as we are used to inside the function, but are not visible outside it. The potential source of confusion is that global variables are also accessible inside a function, just as everywhere else in the code. We could have assigned a value to the variables P and r outside the function, anywhere before the first call to amount, and the code would still work:

```
P = 100
r = 5.0

def amount(n):
    return P*(1+r/100)**n

print(amount(7))
```

Here n is passed as an argument, while, for P and r, the values assigned outside the function is used. However, it is also possible to define local and global variables with the same name, such as

```
P = 100
r = 5.0

def amount(n):
    r = 4.0
    return P*(1+r/100)**n
```

Which value of r is used in the function call here? Local variable names always take precedence over the global names. When the mathematical formula is encountered in the code above, Python will look for the values of the variables P, r, and n that appear in the formula. First, the so-called *local namespace* is searched, that is, Python looks for local variables with the given names. If local variables are found, as for r and n in this case, these values are used. If some variables are not found in the local namespace, Python will move to the *global namespace*, and look for global variables that match the given names. If a variable with the right name is found among the global variables, that is, it has been defined in the main program, then the corresponding value is used. If no global variable with the right name is found there are no more places to search, and the program ends with an error message. This sequential search for variables is quite natural and logical, but still a potential source of confusion and programming errors. Additional confusion can arise if we attempt to change a global variable inside a function. Consider, for instance, this small extension of the code above:

```
P = 100
r = 5.0

def amount(n):
    r = 4.0
    return P*(1+r/100)**n

print(amount(n=6))
print(r)
```

```
126.53190184960003
5.0
```

As revealed by the print statements, r is set to 4.0 inside the function, but the global variable r remains unchanged after the function has been called. Since the line r = 4.0 occurs inside a function, Python will treat this as the definition of a new local variable, rather than trying to change a global one. We thus define a new local r with value 4.0, while there is still another r defined in the global namespace. After the function has ended, the local variable no longer exists (in programming terms, it *goes out of scope*), whereas the global r is still there and has its original value. If we actually want to change a global variable inside a function, we must explicitly state so by using the keyword global. Consider this minor change of the code above:

```
P = 100
r = 5.0

def amount(n):
    global r
    r = 4.0
    return P*(1+r/100)**n

print(amount(n=6))
print(r)
```

```
126.53190184960003
4.0
```

In this case, the global r is changed. The keyword global tells Python that we do want to change a global variable, and not define a new local one. As a general rule, one should minimize the use of global variables inside functions and, instead, define all the variables used inside a function either as local variables or as arguments passed to the function. to the function. Similarly, if we want the function to change a global variable then we should make the function return this variable, instead of using the keyword global. It is difficult to think of a single example where using global is the best solution, and in practice it should never be used. If we actually wanted the function above to change the global r, the following is a better way:

```python
P = 100
r = 5.0

def amount(n,r):
    r = r - 1.0
    a = P*(1+r/100)**n
    return a, r

a0, r = amount(7)
print(a0, r)
```

Notice that, here, we return two values from the function, separated by a comma, just as in the list of arguments, and we also assign the returned values to the global variables a0, r in the line where the function is called. Although this simple example might not be the most useful in practice, there are many cases in which it is useful for a function call to change a global variable. In such cases the change should always be performed in this way, by passing the global variable in as an argument, returning the variable from the function, and then assigning the returned value to the global variable. Following these steps is far better than using the global keyword inside the function, since it ensures that each function is a self-contained entity, with a clearly defined interface to the rest of the code through the list of arguments and return values.

**Multiple return values are returned as a tuple.** For a more practically relevant example of multiple return values, say we want to implement a mathematical function so that both the function value and its derivative are returned. Consider, for instance, the simple physics formula that describes the height of an object in vertical motion; $y(t) = v_0 t + (1/2)gt^2$, where $v_0$ is the initial velocity, $g$ is the gravitational constant, and $t$ is time. The derivative of the function is $y'(t) = v_0 - gt$, and we can implement a Python function that returns both the function value and the derivative:

```python
def yfunc(t, v0):
    g = 9.81
    y = v0*t - 0.5*g*t**2
    dydt = v0 - g*t
    return y, dydt
```

```
# call:
position, velocity = yfunc(0.6, 3)
```

As above, the return arguments are separated by a comma, and we assign the values to the two global variables `position` and `velocity`, also separated by a comma. When a function returns multiple values like this, it actually returns a tuple, the immutable list type defined in the previous chapter. We could therefore replace the call above with something like the following:

```
pos_vel = yfunc(0.6,3)
print(pos_vel)
print(type(pos_vel))
```

```
(0.034199999999999786, -2.886)
<class 'tuple'>
```

We see that the function returns a tuple with two elements. In the previous call, when we included a comma-separated list of variable names on the left-hand side (i.e., `position, velocity`), Python would *unpack* the elements in the tuple into the corresponding variables. For this unpacking to work, the number of variables must match the length of the tuple; otherwise, we obtain an error message stating that there are too many or not enough values to unpack.

A function can return any number of arguments, separated by commas exactly as above. Here we have three:

```
def f(x):
    return x, x**2, x**4

s = f(2)
print(type(s), s)
x, x2, x4 = s
```

Notice the last line, where a tuple of length 3 is unpacked into three individual variables.

**Example: A function to compute a sum.** For a more relevant function example, of a kind that will arise frequently in this book, consider the sum

$$L(x;n) = \sum_{i=1}^{n} \frac{x^i}{i},$$

which is an approximation to $-\ln(1-x)$ for a finite $n$ and $|x| < 1$. The corresponding Python function for $L(x;n)$ looks like

```
def L(x,n):
    s = 0
    for i in range(1,n+1):
        s += x**i/i

    return s
```

```
#example use
x = 0.5
from math import log
print(L(x, 3), L(x, 10), -log(1-x))
```

The output from the print statement indicates that the approximation improves as the number of terms $n$ is increased, as is usual for such approximating series. For many purposes, it would be useful if the function returned the error of the approximation, that is, $-\ln(1-x) - L(x;n)$, in addition to the value of the sum:

```
from math import log

def L2(x, n):
    s = 0
    for i in range(1,n+1):
        s += x**i/i
    value_of_sum = s

    error = -log(1-x) - value_of_sum
    return value_of_sum, error

# typical call:
x = 0.8; n = 10
value, error = L2(x, n)
```

**A function does not need a return statement.** All the functions considered so far have included a return statement. While this will be the case for most of the functions we write in this course, there will be exceptions, and a function does not need to have a return statement. For instance, some functions only serve the purpose of printing information to the screen, as in

```
def somefunc(obj):
    print(obj)

return_value = somefunc(3.4)
```

Here, the last line does not make much sense, although it is actually valid Python code and will run without errors. If somefunc does not return anything, how can we then call the function and assign the result to a variable? If we do not include a return statement in a function, Python will automatically return a variable with value None. The value of the variable return_value in this case will therefore be None, which is not very useful, but serves to illustrate the behavior of a function with no return statement. Most functions we will write in this course will either return variables or print or plot something to the screen. One typical use of a function without a return value is to print information in a tabular format to the screen. This is useful in many contexts, including studying the convergence of series approximations such as the one above. The following function calls the L2(x,n) function defined above, and uses a for loop to print relevant information in a nicely formatted table:

```
def table(x):
    print(f'x={x}, -ln(1-x)={-log(1-x)}')
    for n in [1, 2, 10, 100]:
        value, error = L2(x, n)
        print(f'n={n:4d} approx: {value:7.6f}, error: {error:7.6f}')

table(0.5)
```

```
x=0.5, -ln(1-x)=0.6931471805599453
n=   1 approx: 0.500000, error: 0.193147
n=   2 approx: 0.625000, error: 0.068147
n=  10 approx: 0.693065, error: 0.000082
n= 100 approx: 0.693147, error: 0.000000
```

This function does not need to return anything, since entire purpose is to print information to the screen.

## 4.3 Default Arguments and Doc Strings

When we used the range-function in the previous chapter, we saw that we could vary the number of arguments in the function call from one to three, and the non-specified arguments would be assigned default values. We can achieve the same functionality in our own functions, by defining *default arguments* in the function definition:

```
def somefunc(arg1, arg2, kwarg1=True, kwarg2=0):
    print(arg1, arg2, kwarg1, kwarg2)
```

A function defined in this way can be called with two, three, or four arguments. The first two have no default value and must therefore be included in the call, while the last two are optional and will be set to the default value if not specified in the call. In texts on Python programming, *default arguments* are often referred to as keyword arguments, although these terms do not mean exactly the same thing. They are, however, closely related, which is why the terms are sometimes used interchangeably. Just as we cannot have keyword arguments preceding positional arguments in a function call, we cannot have default arguments preceding non-default arguments in the function header. The following code demonstrates uses of the alternative function calls for a useless but illustrative function. Testing a simple function such as the following, which does nothing but print out the argument values, is a good way to understand the implications of default arguments and the resulting flexibility in argument lists:

```
>>> def somefunc(arg1, arg2, kwarg1=True, kwarg2=0):
>>>     print(arg1, arg2, kwarg1, kwarg2)

>>> somefunc('Hello', [1,2])    # drop kwarg1 and kwarg2
```

```
Hello [1, 2] True 0              # default values are used

>>> somefunc('Hello', [1,2], 'Hi')
Hello [1, 2] Hi 0               # kwarg2 has default value

>>> somefunc('Hello', [1,2], 'Hi', 6)
Hello [1, 2] Hi 0               # kwarg2 has default value

>>> somefunc('Hello', [1,2], kwarg2='Hi') #kwarg2
Hello [1, 2] True Hi           # kwarg1 has default value

>>> somefunc('Hello', [1,2], kwarg2='Hi', kwarg1=6)
Hello [1, 2] 6 Hi             # specify all args
```

Using what we now know about default arguments, we can improve the function considered above, which implements the formula

$$y(t) = v_0 t - \frac{1}{2}gt^2.$$

Here, it could be natural to think of $t$ as the primary argument to the function, which should always be provided, while $v_0$ and possibly also $g$ could be provided as default arguments. The function definition in Python could read

```
def yfunc(t, v0=5, g=9.81):
    y = v0*t - 0.5*g*t**2
    dydt = v0 - g*t
    return y, dydt

#example calls:
y1, dy1 = yfunc(0.2)
y2, dy2 = yfunc(0.2,v0=7.5)
y3, dy3 = yfunc(0.2,7.5,10.0)
```

**Documentation of Python functions.** An important Python convention is to document the purpose of a function, its arguments, and its return values in a *doc string* - a (triple-quoted) string written immediately after the function header. The doc string can be long or short, depending on the complexity of the function and its inputs and outputs. The following two examples show how a doc string can be used:

```
def amount(P, r, n):
    """Compute the growth of an investment over time."""
    a = P*(1+r/100.0)**n
    return a

def line(x0, y0, x1, y1):
    """
    Compute the coefficients a and b in the mathematical
    expression for a straight line y = a*x + b that goes
    through two points (x0, y0) and (x1, y1).

    x0, y0: a point on the line (floats).
```

```
    x1, y1: another point on the line (floats).
    return: a, b (floats) for the line (y=a*x+b).
    """
    a = (y1 - y0)/(x1 - x0)
    b = y0 - a*x0
    return a, b
```

Doc strings do not take much time to write, and are very useful for others who want to use the function. A widely accepted convention in the Python community, doc strings are also used by various tools for automatically generating nicely formatted software documentation. Much of the online documentation of Python libraries and modules is automatically generated from doc strings included in the code.

## 4.4 If-Tests for Branching the Program Flow

In computer programs we often want to perform different actions depending on a condition. As usual, we can find a similar concept in mathematics that should be familiar to most readers of this book. Consider a function defined in a piecewise manner, for instance,

$$f(x) = \begin{cases} \sin x, 0 \leq x \leq \pi \\ 0, \quad \text{otherwise} \end{cases}$$

The Python implementation of such a function needs to test the value of the input $x$, and return either zero or $\sin(x)$ depending on the outcome. Such a decision in the program code is called *branching* and is obtained using an if-test, or, more generally, an if-else block. The code looks like

```
from math import sin, pi

def f(x):
    if 0 <= x <= pi:
        return sin(x)
    else:
        return 0

print(f(0.5))
print(f(5*pi))
```

The new item here is the if-else block. An if-test is simply constructed by the keyword `if` followed by a Boolean variable or expression, and then a block of code which is to be executed if the condition is true. When the if-test is reached in the function above, the Boolean condition is tested, just as for the while loops in the previous chapter. If the condition is true, the following block of indented code is executed (in this case, just one line); if not, the indented code block after `else` is executed. You might also notice that, unlike the functions seen so far, this function has two return statements. This is perfectly valid and is quite common in functions with if-tests. When a return statement is executed, the function call is over and any following lines in the function are simply ignored. Therefore, there is usually no point in having multiple return statements unless they are combined with if-tests, since, if the first one is always executed the others will never be reached.

Sometimes we just want a piece of code to be executed if a condition is true, and to do nothing otherwise. In such cases, we can skip the `else` block and define only an if-test:

```
if condition:
    <block of statements, executed if condition is True>

<next line after if-block, always executed>
```

Here, whatever is inside the if-block is executed if `condition` is true, otherwise the program simply moves to the next line after the block. As above, we can add an else-block to ensure that exactly one of two code blocks is executed

```
if condition:
    <block of statements, executed if condition is True>
else:
    <block of statements, executed if condition is False>
```

For mathematical functions of the form considered above we usually want to include an else-block, since we want the function to return a meaningful value for all input arguments. Forgetting the else-block in the definition `f(x)` above would make the function return $\sin(x)$ (a `float`) for $0 \leq x \leq \pi$, and otherwise `None`, which is obviously not what we want. Finally, we cans combine multiple if-else statements with different conditions

```
if condition1:
    <block of statements>
elif condition2:
    <block of statements>
elif condition3:
    <block of statements>
else:
    <block of statements>
<next statement>
```

Notice the keyword `elif`, short for *else if*, which ensures that that subsequent conditions are only tested only if the preceding ones are False. The conditions are checked one by one and, as soon as one is evaluated as true, the corresponding block is executed and the program moves to the first statement after the else block. The remaining conditions are not checked. If none of the conditions is true, the code inside the else block is executed.

Multiple branching has useful applications in mathematics, since we often see piecewise functions defined on multiple intervals. Consider for instance the piecewise linear function

$$N(x) = \begin{cases} 0, & x < 0 \\ x, & 0 \le x < 1 \\ 2 - x, & 1 \le x < 2 \\ 0, & x \ge 2 \end{cases}.$$

which in Python can be implemented with multiple if-else-branching

```
def N(x):
    if x < 0:
        return 0
    elif 0 <= x < 1:
        return x
    elif 1 <= x < 2:
        return 2 - x
    elif x >= 2:
        return 0
```

In later chapters we will see multiple examples of more general use of branching, not restricted to mathematics or piecewise-defined functions.

**Inline if-tests for shorter code.** The list comprehensions in Chapter 3 offered a more compact alternative to the standard way of defining lists, and a similar alternative exists for if-tests. A common use of if-else blocks is to assign a value to a variable, where the value depends on some condition, just as in the examples above. The general form looks like

```
if condition:
    variable = value1
else:
    variable = value2
```

This code can be replaced by the following one-line if-else block:

```
variable = (value1 if condition else value2)
```

Using this compact notation, we can write the example from the start of this section as

```
def f(x):
    return (sin(x) if 0 <= x <= pi else 0)
```

## 4.5 Functions as Arguments to Functions

Arguments to Python functions can be any Python object, including another function. This functionality is quite useful for many scientific applications, where we need to define mathematical functions that operate on or make use of other mathematical functions. For instance, we can easily write Python functions for numerical approximations of integrals $\int_a^b f(x)dx$, derivatives $f'(x)$, and roots $f(x) = 0$. For such functions to be general and useful, they should work with an arbitrary $f(x)$, which is most conveniently accomplished by passing a Python function `f(x)` as an argument to the function.

Consider the example of approximating the second derivative $f''(x)$ by centered finite differences,

$$f''(x) \approx \frac{f(x-h) - 2f(x) + f(x+h)}{h^2}.$$

The corresponding Python function looks like

```
def diff2(f, x, h=1E-6):
    r = (f(x-h) - 2*f(x) + f(x+h))/float(h*h)
    return r
```

We see that the function `f` is passed to the function just as any other argument, and is called as a regular function inside `diff2`. Of course, for this to work, we need to actually send a callable function as the first argument to `diff2`. If we send something else, like a number or a string, the code will stop with an error when it tries to make the call `f(x-h)` in the next line. Such potential errors are part of the price we pay for Python's flexibility. We can pass any argument to a function, but the object we pass must be possible to use as intended inside the function. As noted above, for more complex functions, it is useful to include a doc string that specifies the types of arguments the function expects.

**Lambda functions for compact inline function definitions.** In order to use the function `diff2` above, one would standardly define our `f(x)` as a Python function, and then pass it as an argument to `diff2`. The following code shows an example:

```
def f(x):
    return x**2 - 1

df2 = diff2(f,1.5)
print(df2)
```

The concept known as a *lambda function* offers a compact way to define functions, which can be convenient for the present application. Using the keyword `lambda`, we can define our `f` on a single line, as follows:

```
f = lambda x: x**2 - 1
```

More generally, a lambda function defined by

```
somefunc = lambda a1, a2, ...: some_expression
```

is equivalent to

```
def somefunc(a1, a2, ...):
    return some_expression
```

It could be natural to ask whether anything is really gained here, and whether it is useful to introduce a new concept just to reduce a function definition

from two lines to one line. One answer is that the lambda function definition
can be placed directly in the argument list of the other function. Instead of
first defining f(x) and then passing it as an argument, as in the code above,
we can combine these tasks into one line:

```
df2 = diff2(lambda x: x**2-1,1.5)
print(df2)
```

Using lambda functions in this way can be quite convenient in cases in which
we need to pass a simple mathematical expression as an argument to a Python
function. We save some typing, and could also improve the code's readability.


## 4.6 Solving Equations with Python Functions

Solving equations of the form $f(x) = 0$ is a frequently occuring task in all
branches of science and engineering. For special cases, such as a linear or
quadratic $f$, we have simple formulas that give us the solution directly. In
the general case, however, the equation cannot be solved analytically, and we
need to find an approximate solution using numerical methods. We shall see
that we can create powerful and flexible tools for equation solving based on
the building blocks introduced so far. Specifically, we will combine functions
and function arguments with the while loop introduced in Chapter 3.

**Finding roots on an interval with the bisection method.** One of the
simplest algorithms for solving equations of the form $f(x) = 0$ is called the
*bisection method.* This method is founded on the intermediate value theorem,
which states that, if a continuous function changes sign on an interval $[a,b]$
then there must be a value $x \in [a,b]$ such that $f(x) = 0$. In the bisection
method we start by choosing an interval $[a,b]$ on which $f$ changes sign (i.e.,
$f(a)f(b) < 0$), and then compute the midpoint $m = (a+b)/2$ and check the
sign of $f(m)$. If $f$ changes sign on $[a,m]$ then we repeat the process on the
interval $[a,m]$; otherwise, we choose $[m,b]$ as our new interval and repeat the
process there. These steps are conveniently implemented as a while loop, and
we can create a generic tool by placing the while loop inside a function that
takes a function as argument:

```
from math import exp

def bisection(f,a,b,tol= 1e-3):
    if f(a)*f(b) > 0:
        print(f'No roots or more than one root in [{a},{b}]')
        return

    m = (a+b)/2

    while abs(f(m)) > tol:
        if f(a)*f(m) < 0:
```

```
            b = m
        else:
            a = m
        m = (a+b)/2
    return m

#call the method for f(x)= x**2-4*x+exp(-x)
f = lambda x: x**2-4*x+exp(-x)
sol = bisection(f,-0.5,1,1e-6)

print(f'x = {sol:g} is an approximate root, f({sol:g}) = {f(sol):g}')
```

We see that the `bisection` function takes four arguments: the mathematical function $f(x)$ implemented as a Python function, the bounds for our initial interval, and the tolerance for the approximate solution. The first if-test of the function simply checks that $f$ changes sign in $[a, b]$, which ensures that the function has at least one root on the interval. We then proceed to define the midpoint `m` and enter the while-loop, which forms the core of the algorithm. This loop will continue running as long as `abs(f(m)) > tol` (otherwise `m` is our solution), repeatedly checking whether $f$ changes sign on $[a, m]$ or $[m, b]$, and then calculating a new `m` to repeat the process on an interval of half the size.

**Newton's method gives faster convergence.** The bisection method converges quite slowly, and other methods are far more popular for solving non-linear equations. In particular, numerous varieties of Newton's method are widely used in practice. Newton's method is based on a *local linearization* of the non-linear function $f(x)$. Starting with an initial guess $x_0$, we replaces $f(x)$ by a linear function $g(x)$ that satisfies $g(x) \approx f(x)$ in a small interval around $x_0$. Then, we solve the equation $g(x) - 0$ to find an updated guess $x_1$, and repeat the process of linearization around that point. Repeated application of these steps converges quickly towards the true solution, provided that the initial guess $x_0$ is sufficiently close. In mathematics, one step of the algorithm looks like

$$x_{n+1} = x_n - \frac{f(x_n)}{f'(x_n)},$$

where $x_n$ is the solution after $n$ iterations, $x_{n+1}$ is the improved approximation, and $f'(x_n)$ is the derivative of $f$ in $x_n$.

Just as the bisection method, Newton's method is easy to implement in a while loop, and we can implement it as a generic function that takes a Python function implementing $f(x)$ as argument. The function will also need $f'(x)$, since this is used in the algorithm, as well as an initial guess $x_0$ and a tolerance:

```
from math import exp

def Newton(f, dfdx, x0, tol= 1e-3):
    f0 = f(x0)
    while abs(f0) > tol:
```

```
            x1 = x0 - f0/dfdx(x0)
            x0 = x1
            f0 = f(x0)
        return x0

#call the method for f(x)= x**2-4*x+exp(-x)
f = lambda x: x**2-4*x+exp(-x)
dfdx = lambda x: 2*x-4-exp(-x)


sol = Newton(f,dfdx,0,1e-6)

print(f'x = {sol:g} is an approximate root, f({sol:g}) = {f(sol):g}')
```

Notice how the x0 variable is updated inside the loop. The algorithm only
needs to know the value at one iteration to compute the next one, so for each
iteration we update x0 to hold the most recent approximation, and then use
this to compute the next one. Note also that the implementation provided
here is not very robust, and if the method does not converge, it will simply
continue in an infinite loop. One simple way to improve the implementation
is to stop the method after a given number of iterations:

```
from math import exp

def Newton2(f, dfdx, x0, max_it=20, tol= 1e-3):
    f0 = f(x0)
    iter = 0
    while abs(f0) > tol and iter < max_it:
        x1 = x0 - f0/dfdx(x0)
        x0 = x1
        f0 = f(x0)
        iter += 1

    converged = iter < max_it
    return x0, converged, iter

#call the method for f(x)= x**2-4*x+exp(-x)
f = lambda x: x**2-4*x+exp(-x)
dfdx = lambda x: 2*x-4-exp(-x)

sol, converged, iter = Newton2(f,dfdx,0,tol=1e-3)

if converged:
    print(f'Newtons method converged in {iter} iterations')
else:
    print(f'The method did not converge')
```

Newton's method usually converges much faster than the bisection method,
but has the disadvantage the function $f$ needs to be manually differentiated.
In Chapter 8 we will see some examples of how this step can be avoided.

# 4.7 Writing Test Functions to Verify our Programs

In the first part of this chapter, we mentioned the idea of writing tests to verify that functions work as intended. This approach to programming can be very effective, and although we spend some time writing the tests, we often save much more time by the fact that we discover errors early, and can build our program from components that are known to work. The process is often referred to as *unit testing*, since each test verifies that a small unit of the program works as expected. Many programmers even take the approach one step further and write the test before they write the actual function. This approach is often referred to as test-driven development and is an increasingly popular method for software development.

The tests we write to test our functions are also functions, a special type of function known as *test functions*. Writing good test functions, which test the functionality of our code in a reliable manner, can be quite challenging; however, the overall idea of test functions is very simple. For a given function, which often takes one or more arguments, we choose arguments such that we can calculate the result of the function by hand. Inside the test function, we then simply call our function with the right arguments and compare the result returned by the function with the expected (hand-calculated) result. The following example illustrates how we can write a test function to test that the (very) simple function `double(x)` works as it should:

```python
def double(x):         # some function
    return 2*x

def test_double():     # associated test function
    x = 4              # some chosen x value
    expected = 8       # expected result from double(x)
    computed = double(x)
    success = computed == expected  # Boolean value: test passed?
    msg = f'computed {computed}, expected {expected}'
    assert success, msg
```

In this code, the only Python keyword that we have not seen previously is `assert`, which is used instead of `return` whenever we write a test function. Test functions should not return anything, so a regular return statement would not make sense. The only purpose of the test function is to compare the value returned by a function with the value we expect it to return, and to write an error message if the two are different. This task is precisely what `assert` does. The keyword `assert` should always be followed by a condition, `success` in the code above, that is true if the test passes and false if it fails. The code above follows the typical recipe; we compare the expected with the returned result in `computed == expected`, which is a Boolean expression returning true or false. This value is then assigned to the variable `success`, which is included in the `assert` statement. The last part of the assert statement, the text string `msg`, is optional and is simply included to give a more meaningful error message if the test fails. If we leave this out, and only write `assert success`, we will see a general message stating that the test has failed (a so-called *assertion error*), but without much information about what actually went wrong.

Some rules should be observed when writing test functions:

- The test function must have at least one statement of the type `assert success`, where `success` is a Boolean variable or expression, which is true if the test passed and false otherwise. We can include more than one assert statement if we want, but we always need at least one.
- The test function should take no arguments. The function to be tested will typically be called with one or more arguments, but these should be defined as local variables inside the test function.
- The name of the function should always be `test_`, followed by the name of the function we want to test. Following this convention is useful because it makes it obvious to anyone reading the code that the function is a test function, and it is also used by tools that can automatically run all test functions in a given file or directory. More about this is discussed below.

If we follow these rules, and remember the fundamental idea that a test function simply compares the returned result with the expected result, writing test functions does not have to be complicated. In particular, many of the functions we write in this course will evaluate some kind of mathematical function and then return either a number or a list/tuple of numbers. For this type of function, the recipe for test functions is quite rigid, and the structure is usually exactly the same as in the simple example above.

If you are new to programming, it can be confusing to be faced with a general task such as "write a test function for the Python function `somefunc(x,y)`," and it is natural to ask questions about what arguments the function should be tested for and how you can know what the expected values are. In such cases it is important to remember the overall idea of test functions, and also that these are choices that must be made by the programmer. You have to choose a set of suitable arguments, then calculate or otherwise predict by hand what the function *should* return for these arguments, and write the comparison in the test function.

**A test function can include multiple tests.** We can have multiple assert statements in a single test function. This can be useful if we want to test a function with different arguments. For instance, if we write a test function for one of the piecewise-defined mathematical functions considered earlier in this chapter, it would be natural to test all the separate intervals on which the function is defined. The following code illustrates how this can be done:

```
from math import sin, pi

def f(x):
    if 0 <= x <= pi:
```

```
            return sin(x)
        else:
            return 0

def test_f():
    x1, exp1 = -1.0, 0.0
    x2, exp2 = pi/2, 1.0
    x3, exp3 = 3.5, 0.0

    tol = 1e-10
    assert abs(f(x1)-exp1) < tol, f'Failed for x = {x1}'
    assert abs(f(x2)-exp2) < tol, f'Failed for x = {x2}'
    assert abs(f(x3)-exp3) < tol, f'Failed for x = {x3}'
```

Note here that, since we compare floating point numbers, which have finite precision on a computer, we compare with a tolerance rather than the equality ==. The tolerance `tol` is some small number, chosen by the programmer, that is small enough that we would consider a difference of this magnitude insignificant, but greater than the machine precision ($\approx 10^{-16}$). In practice, comparing floats using == will quite often work, but sometimes it fails and it is impossible to predict when this will happen. The code therefore becomes unreliable, and it is much safer to compare with a tolerance. On the other hand, when we work with integers , we can always use ==.

One could argue that the test function code above is quite inelegant and repetitive, since we repeat the same lines multiple times with very minor changes. Since we only repeat three lines, it might not be a big deal in this case, but if we included more assert statements it would certainly be both boring and error-prone to write code in this way. In the previous chapter, we introduced loops as a much more elegant tool f or performing such repetitive tasks. Using lists and a for loop, the example above can be written as follows:

```
from math import sin, pi

def f(x):
    if 0 <= x <= pi:
        return sin(x)
    else:
        return 0

def test_f():
    x_vals = [-1, pi/2, 3.5]
    exp_vals = [0.0, 1.0, 0.0]
    tol = 1e-10
    for x, exp in zip(x_vals, exp_vals):
        assert abs(f(x)-exp) < tol, \
            f'Failed for x = {x}, expected {exp}, but got {f(x)}'
```

**Python tools for automatic testing.** An advantage of following the naming convention for test functions defined above is that there are tools that can be used to automatically run *all* the test functions in a file or folder and report if any bug has sneaked into the code. The use of such automatic testing tools is essential in larger development projects with multiple people working on the same code, but can also be quite useful for your own projects. The recommended and most widely used tool is called `pytest` or `py.test`, where `pytest` is simply the new name for `py.test`. We can run `pytest` from the terminal window, and pass it either a file name or a folder name as an argument, as in

```Terminal
Terminal> pytest .
Terminal> pytest my_python_project.py
```

If we pass it a file name, `pytest` will look for functions in this file with a name starting with `test_`, as specified by the naming convention above. All these functions will be identified as test functions and called by `pytest`, regardless of whether the test functions are actually called from elsewhere in the code. After execution, `pytest` will print a short summary of how many tests it found, and how many that passed and failed.

For larger software projects, it might be more relevant to give a directory name as argument to `pytest`, as in the first line above. In this case, the tool will search the given directory (here ., the directory we are currently in) and all its sub-directories for Python files with names starting or ending with `test` (e.g., `test_math.py`, `math_test.py`, etc.). All these files will be searched for test functions following the naming convention, and these will be run as above. Large software projects typically have thousands of test functions, and it is very convenient to collect them in a separate file and use automatic tools such as `pytest`. For the smaller programs we write in this course, it can be just as easy to write the test functions in the same file as the functions being tested.

It is important to remember that test functions run *silently* if the test passes; that is, we only obtain an output if there is an assertion error, otherwise nothing is printed to the screen. When using `pytest` we are always given a summary specifying how many tests were run, but if we include calls to the test functions directly in the `.py` file, and run this file as normal, there will be no output if the test passes. This can be confusing, and one is sometimes left wondering if the test was called at all. When first writing a test function, it can be useful to include a print-statement inside the function, simply to verify that the function is actually called. This statement should be removed once we know the function works correctly and as we become used to how the test functions work.

# Chapter 5
# User Input and Error Handling

So far, all the values we have assigned to variables have been written directly into our programs. If we want a different value of a variable, we need to edit the code and rerun the program. Of course, this is not how we are used to interacting with computer programs. Usually, a program will receive some input from users, most often through a graphical user interface (GUI). However, although GUIs dominate in modern human–computer interaction, other ways of interacting with computer programs can be just as efficient and, in some cases, far more suitable for processing large amounts of data and automating repetitive tasks. In this chapter we will show how we can extend our programs with simple yet powerful systems for user input. In particular, we will see how a program can receive command line arguments when it is run, how to make a program stop and ask for user input, and how a program can read data from files.

A side effect of allowing users to interact with programs is that things will often go wrong. Users will often provide the wrong input, and programs should be able to handle such events without simply stopping and writing a cryptic error message to the screen. We will introduce a concept known as *exception handling*, which is a widespread system for handling errors in programs, used in Python and many other programming languages.

Finally, in this chapter, we shall see how to create our own modules that can be imported for use in other programs, just as we have done with the `math` module in previous chapters.

## 5.1 Reading User Input Data

So far, we have implemented various mathematical formulas that involved input variables and parameters, but all of these values have been hard-coded into the programs. To introduce a new example, consider the following formula, which gives an estimate of the atmospheric pressure $p$ as a function of

© The Author(s) 2020
J. Sundnes, *Introduction to Scientific Programming with Python*, Simula SpringerBriefs on Computing 6,
https://doi.org/10.1007/978-3-030-50356-7_5

altitude $h$:
$$p = p_0 e^{-h/h_0},$$

where $p_0$ is the pressure at sea level ($\approx 100$ kPa) and $h_0$ is the so-called scale height ($\approx 8.4$km). A Python program for evaluating this formula could look like

```
from math import exp

p0 = 100.0      #sea level pressure (kPa)
h0 = 8400       #scale height (m)

h = 8848
p = p0 * exp(-h/h0)
print(p)
```

Of course, we are usually interested in evaluating the formula for different altitudes, which, in this code, would require editing the line h = 8848 to change the respective variable, and then rerunning the program. This solution could be acceptable for programs we write and use ourselves, but it is not how we are used to interacting with computers. In particular, if we write programs that could be used by others, editing the code this way is inconvenient and can easily introduce errors.

For our programs to be robust and usable, they need to be able to read relevant input data from the user. We will consider three different ways to accomplish this, each with its strengths and weaknesses. We will (i) create programs that stop and ask for user input, and then continue the execution when the input is received; (ii) enable our programs to receive *command line arguments*, that is, arguments provided when we run the program from the terminal; and (iii) make the programs read input data from files.

**Obtaining input from questions and answers.** A natural extension of this program is to allow it to ask the user for a value of h, and then compute and output the corresponding atmospheric pressure. A Python function called input provides exactly this functionality. For instance a line such as

```
input('Input the altitude (in meters):')
```

will make the program stop and display the text Input the altitude (in meters): in the terminal, and then continue when the user presses Enter. The complete code could look like

```
from math import exp

h = input('Input the altitude (in meters):')
h = float(h)

p0 = 100.0      #sea level pressure (kPa)
h0 = 8400       #scale height (m)

p = p0 * exp(-h/h0)
print(p)
```

Running the program in a terminal window could look like:

```
Terminal> python altitude.py
Input the altitude (in meters): 2469
74.53297273796525
```

Notice in particular the line h = float(h), which is an example of the type conversions mentioned in Chapter 2. The input function will always return a text string, which must be converted to an actual number before we can use it in computations. Forgetting this line in the code above will lead to an error in the line that calculates amount, since we would by trying to multiply a string with a float. From these considerations, we can also imagine how easy it is to break the program above. The user can type any string, or simply press enter (which makes h an empty string), but the conversion h = float(h) only works if the string is a number.

As another example, consider a program that asks the user for an integer n and prints the n first even numbers:

```
n = int(input('n=? '))

for i in range(1, n+1):
    print(2*i)
```

Here we convert the input text using int(...), since the range function only accepts integer arguments. Just as in the example above, the code is not very robust, since it will break from any input that cannot be converted to an integer. Later in this chapter we will look at ways to handle such errors and make the programs more robust.

**Command line arguments are words written after the program name.** When working in a Unix-style terminal window (e.g., Mac, Linux, Windows PowerShell), we often provide arguments when we run a command. These arguments can be names of files or directories, for example, when copying a file with cp, or they can change the output from the command, such as ls -l to obtain more detailed output from the ls command. Anyone who is used to working in Unix-style terminals will be familiar with commands like these:

```
Terminal> cp -r yourdir ../mydir
Terminal> ls -l
terminal> cd ../mydir
```

Some commands require arguments – for instance, you receive an error message if you do not give two arguments to cp – while other arguments are optional. Standard Unix programs make heavy use of command line arguments, (try, for instance, typing man ls), because they are a very efficient

way of providing input and modifying program behavior. We will make our
Python programs do the same, and write programs that can be run as

---
Terminal
---
```
Terminal> python myprog.py arg1 arg2 arg3 ...
```
---

where `arg1 arg2 arg3`, and so forth are input arguments to the program.

We again consider the air pressure calculation program above, but now we
want the altitude to be specified as a command line argument rather than
obtained by stopping and asking for input. For instance, we want to run the
program as followss:

---
Terminal
---
```
Terminal> python altitude_cml.py 2469
74.53297273796525
```
---

To use command line arguments in a Python program, we need to import
a module named `sys`. More specifically, the command line arguments, or,
in reality, any words we type after the command `python altitude.py`, are
automatically stored in a list named `sys.argv` (short for argument values)
and can be accessed from there:

```python
import sys
from math import exp

h = sys.argv[1]
h = float(h)

p0 = 100.0      #sea level pressure (kPa)
h0 = 8400       #scale height (m)

p = p0 * exp(-h/h0)
print(p)
```

Here, we see that we pull out the element with index one from the `sys.argv`
list, and convert it to a float. Just as the input provided with the `input`
function above, the command line arguments are always strings and need
to be converted to floats or integers before they are used in computations.
The `sys.argv` variable is simply a list that is created automatically when
your Python program is run. The first element, `sys.argv[0]` is the name
of the .py-file containing the program. The remainder of the list is made
up of whatever words we type after the program filename. Words separated
by a space become separate elements in the list. A nice way to gain a feel
for the use of `sys.argv` is to test a simple program that will just print out
the contents of the list, for instance, by writing this simple code into the file
`print_cml.py`:

```python
import sys
print(sys.argv)
```

Running this program in different ways illustrates how the list works; for instance,

---
Terminal

---

```
Terminal> python print_cml.py 21 string with blanks 1.3
['print_cml.py', '21', 'string', 'with', 'blanks', '1.3']

Terminal> python print_cml.py 21 "string with blanks" 1.3
['print_cml.py', '21', 'string with blanks', '1.3']
```

---

We see from the second example that, if we want to read in a string containing blanks as a single command line argument, we need to use quotation marks to override the default behavior of each word being treated as a separate list element.

## 5.2 Flexible User Input with `eval` and `exec`

Generally, the safest way to handle input data in the form of text strings is to convert it to the specific variable type needed in the program. We did this above, using the type conversions `int(...)` and `float(...)`, and we will see below how such conversions can be made failproof and handle imporper user input. However, Python also offers a couple of more flexible functions to handle input data, namely, `eval` and `exec`, which are nice to know about. Extensive use of these functions is not recommended, especially not in larger programs, since the code can become messy and error-prone. However, they offer some flexible and fun opportunities for handling input data. Starting with `eval`, this function simply takes a string s as input and evaluates it as a regular Python expression, just as if it were written directly into the program. Of course, s must be a legal Python expression, otherwise the code stops with an error message. The following interactive Python session illustrates how `eval` works:

```
>>> s = '1+2'
>>> r = eval(s)
>>> r
3
>>> type(r)
<type 'int'>

>>> r = eval('[1, 6, 7.5] + [1, 2]')
>>> r
[1, 6, 7.5, 1, 2]
>>> type(r)
<type 'list'>
```

Here, the line r = eval(s) is equivalent to writing r = 1+2, but using eval gives much more flexibility, of course, since the string is stored in a variable and can be read as input.

A small Python program using eval can be quite flexible. Consider, for instance, the following code

```
i1 = eval(input('operand 1: '))
i2 = eval(input('operand 2: '))
r = i1 + i2
print(f'{type(i1)} + {type(i2)} becomes {type(r)} with value{r}')
```

This code can handle multiple input types. If we save the code in a file add_input.py and run it from the terminal, we can, for instance, add integer and float numbers, as in:

```
———————————————————————————— Terminal ————————————————————————————
Terminal> python add_input.py
operand 1: 1
operand 2: 3.0
<type 'int'> + <type 'float'> becomes <type 'float'>
with value 4
```

or two lists, as follows:

```
———————————————————————————— Terminal ————————————————————————————
Terminal> python add_input.py
operand 1: [1,2]
operand 2: [-1,0,1]
<type 'list'> + <type 'list'> becomes <type 'list'>
with value [1, 2, -1, 0, 1]
```

We could achieve similar flexibility with conventional type conversion, that is, using float(i1), int(i1), and so on, but that would require much more programming to correctly process the input strings. The eval function makes such flexible input handling extremely compact and efficient, but it also quickly breaks if the input is slightly wrong. Consider the following examples:

```
———————————————————————————— Terminal ————————————————————————————
Terminal> python add_input.py
operand 1: (1,2)
operand 2: [3,4]
Traceback (most recent call last):
  File "add_input.py", line 3, in <module>
    r = i1 + i2
TypeError: can only concatenate tuple (not "list") to tuple

Terminal> python add_input.py
```

```
operand 1: one
Traceback (most recent call last):
  File "add_input.py", line 1, in <module>
    i1 = eval(input('operand 1: '))
  File "<string>", line 1, in <module>
NameError: name 'one' is not defined
```

In the first of these examples, we try to add a tuple and a list, which one could easily imagine would work, but Python does not allow this and therefore the program breaks. In the second example, we try to make the program add two strings, which usually works fine; for instance `"one" +"one"` becomes the string `"oneone"`. However, the eval function breaks when we try to input the first string. To understand why, we need to think about what the corresponding line really means. We try to make the assignment `i1 = eval('one')`, which is equivalent to writing `i1 = one`, but this line does not work unless we have already defined a variable named one. A remedy to this problem is to input the strings with quotation marks, as in the following example

```
                              ┌─────────┐
──────────────────────────────│ Terminal │──────────────────────────────
                              └─────────┘
Terminal> python add_input.py
operand 1: "one"
operand 2: "two"
<class 'str'> + <class 'str'> becomes <class 'str'>
with value onetwo
```

These examples illustrate the benefits of the eval function, and also how it easily breaks programs and is generally not recommended for "real programs". It is useful for quick prototypes, but should usually be avoided in programs that we expect others to use or that we expect to use ourselves over a longer time frame.

The other "magic" text handling function is named exec, and it is fairly similar to eval. However, whereas eval evaluates an *expression*, exec executes a string argument as one or more complete statements. For instance, if we define a string `s = "r = 1+1"`, `eval(s)` is illegal, since the value of s (`"r = 1+1"`) is a statement (an assignment), and not a Python expression. However, `exec(s)` will work fine and is the same as including the line `r = 1+1` directly in the code. The following code illustrates the difference:

```
expression = '1+1'        #store expression in a string
statement = 'r = 1+1'     # store statement in a string
q = eval(expression)
exec(statement)

print(q,r)                # results are the same
```

We can also use `exec` to execute multiple statements, for instance using multi-line strings:

```
somecode = """
def f(t):
    term1 = exp(-a*t)*sin(w1*x)
    term2 = 2*sin(w2*x)
    return term1 + term2
"""
exec(somecode)  # execute the string as Python code
```

Here, the `exec` line will simply execute the string `somecode`, just as if we had typed the code directly in our program. After the call to `exec` we have defined the function `f(t)` and can call this function in the usual way. Although this example does not seem very useful, the flexibility of `exec` becomes more apparent if we combine it with actual user input. For instance, consider the following code, which asks the user to type a mathematical expression involving $x$ and then embeds this expression in a Python function:

```
formula = input('Write a formula involving x: ')
code = f"""
def f(x):
    return {formula}
"""
from math import *  # make sure we have sin, cos, log, etc.
exec(code)          # turn string formula into live function


#Now the function is defined, and we can ask the
#user for x values and evaluate f(x)
x = 0
while x is not None:
    x = eval(input('Give x (None to quit): '))
    if x is not None:
        y = f(x)
        print(f'f({x})={y}')
```

While the program is running, the user is first asked to type a formula, which becomes a function. Then the user is asked to input x values until the answer is `None`, and the program evaluates the function `f(x)` for each x. The program works even if the programmer knows nothing about the user's choice of `f(x)` when the program is written, which demonstrates the flexibility offered by the `exec` and `eval` functions.

To consider another example, say, we want to create a program `diff.py` that evaluates the numerical derivative of a mathematical expression $f(x)$ for a given value of $x$. The mathematical expression and the $x$ value will be given as command line arguments. The program could be used as follows:

```
                          Terminal
Terminal> python diff.py 'exp(x)*sin(x)' 3.4
Numerical derivative: -36.6262969164
```

The derivative of a function $f(x)$ can be approximated with a centered finite difference:

$$f'(x) \approx \frac{f(x+h) - f(x-h)}{2h},$$

for some small $h$. The implementation of the `diff.py` program could look like

```
from math import *
import sys

formula = sys.argv[1]
code = f"""
def f(x):
    return {formula}
"""

exec(code)
x = float(sys.argv[2])

def numerical_derivative(f, x, h=1E-5):
    return (f(x+h) - f(x-h))/(2*h)

print(f'Numerical derivative: {numerical_derivative(f, x)}')
```

Again we see that the flexibility of the `exec` function enables us to implement fairly advanced functionality in a very compact program.

## 5.3 Reading Data from Files

Scientific data are often available in files, and reading and processing data from files have always been important tasks in programming. The data science revolution that we have witnessed in recent years has only increased their importance further, since all data analysis starts with being able to read data from files and store them in suitable data structures. To start with a simple example, consider a file named `data.txt` containing a single column of numbers:

```
21.8
18.1
19
23
26
17.8
```

We assume that we know in advance that there is one number per line, but we do not know the number of lines. How can we read these numbers into a Python program?

The basic way to read a file in Python is to use the function open, which takes a file name as an argument. The following code illustrates its use:

```
infile = open('data.txt', 'r')        # open file
for line in infile:
    # do something with line
infile.close()                         # close file
```

Here, the first line opens the file data.txt for reading, as specified with the letter r, and creates a file object named infile. If we want to open a file for writing, which we will consider later, we have to use open('data.txt','w'). The default is r, so, to read a file we could also simply write infile = open('data.txt'). However, including the r can be a good habit, since it makes the purpose of the line more obvious to anyone reading the code. In the second line, we enter a regular for loop, which will treat the object infile as a list-like object and step through the file line by line. For each pass through the for loop, a single line of the file is read and stored in the string variable line, and inside the for loop we add any code we want for processing this line. When there are no more lines in the file, the for loop ends, just as when looping over a regular list. The final line, infile.close(), closes the file and makes it unavailable for further reading. This line is not very important when reading from files, but it is a good habit to always include it, since it can make a difference when writing to files.

To return to the concrete data file above, say the only processing we want is to compute the mean value of the numbers in the file. The complete code could look like this:

```
infile = open('data.txt', 'r')        # open file
mean = 0
lines = 0
for line in infile:
    number = float(line)               # line is string
    mean = mean + number
    lines += 1
mean = mean/lines
print(f'The mean value is {mean}')
```

This is a standard way to read files in Python, but, as usual, in programming there are multiple ways to do things. An alternative way of opening a file, which many will consider more modern, is by using the following code:

```
with open('data.txt', 'r') as infile:   # open file
    for line in infile:
        # do something with line
```

The first line, using with and as probably does not look familiar, but it does essentially the same thing as the line infile = open(...) in the first example. One important difference is that, if we use with we see that all file reading and processing must be put inside an indented block of code, and the file is automatically closed when this block has been completed. Therefore, the use of `with` to open files is quite popular, and you are likely to see it in Python programs you encounter. The keyword `with` has other uses in Python that we will not cover in this book, but it is particularly common and convenient for reading files and therefore worth mentioning here.

To actually read a file after it has been opened, there are a couple of alternatives to the approach above. For instance, we can read all the lines into a list of strings (lines) and then process the list items one by one:

```
lines = infile.readlines()
infile.close()
for line in lines:
    # process line
```

This approach is very similar to the one used above, but here we are done working directly with the file after the first line, and the for loop instead traverses the list of strings. In practice there is not much difference. Usually, processing files line by line is very convenient, and our good friend the for loop makes such processing quite easy. However, for files with no natural line structure, it can sometimes be easier to read the entire text file into a single string:

```
text = infile.read()
# process the string text
```

The `data.txt` file above contain a single number for each line, which is usually not the case. More often, each line contains many data items, typically both text and numbers, and we might want to treat each one differently. For this purpose Python's string type has a built-in method named `split` that is extremely useful. Say we define a string variable s with some words separated by blank spaces. Then, calling `s.split()` will simply return a list containing the individual words in the string. By default, the words are assumed to be separated by blanks, but if we want a different separator, we can pass it as an argument to `split`. The following code gives some examples:

```
s = "This is a typical string"
csvline = "Excel;sheets;often;use;semicolon;as;separator"
print(s.split())
print(csvline.split())
print(csvline.split(';'))
```

```
['This', 'is', 'a', 'typical', 'string']
['Excel;sheets;often;use;semicolon;as;separator']
['Excel', 'sheets', 'often', 'use', 'semicolon', 'as', 'separator']
```

We see that the first attempt to split the string `csvline` does not work very well, since the string contains no spaces and the result is therefore a list of length one. Specifying the correct separator, as in the last line, solves the problem.

To illustrate the use of `split` in the context of file data, assume we have a file with data on rainfall:

```
Average rainfall (in mm) in Rome: 1188 months between 1782 and 1970
Jan  81.2
Feb  63.2
Mar  70.3
Apr  55.7
May  53.0
Jun  36.4
Jul  17.5
Aug  27.5
Sep  60.9
Oct  117.7
Nov  111.0
Dec  97.9
Year 792.9
```

Although this data file is very small, it is a fairly typical example. Often, there are one or more header lines with information that we are not really interested in processing, and the remainder of the lines contain a mix of text and numbers. How can we read such a file? The key to processing each line is to use split to separate the two words and, for instance, store them in two separate lists for later processing:

```
months = []
values = []
for line in infile:
    words = line.split()  # split into words
    months.append(words[0])
    values.append(float(words[1]))
```

These steps, involving a for loop and then split to process each line, will be the fundamental recipe for all file processing throughout this book. It is important to understand these steps properly and well worth spending some time reading small data files and playing around with split to become familiar with its use. To write the complete program for reading the rainfall data, we must also account for the header line and the fact that the last line contains data of a different type. The complete code could look like:

```
def extract_data(filename):
    infile = open(filename, 'r')
    infile.readline() # skip the first line
    months = []
    rainfall = []
    for line in infile:
        words = line.split() #words[0]: month, words[1]: rainfall
        months.append(words[0])
        rainfall.append(float(words[1]))
    infile.close()
    months = months[:-1]      # Drop the "Year" entry
    annual_avg = rainfall[-1] # Store the annual average
    rainfall = rainfall[:-1]  # Redefine to contain monthly data
    return months, rainfall, annual_avg

months, values, avg = extract_data('rainfall.txt')
```

```
print('The average rainfall for the months:')
for month, value in zip(months, values):
    print(month, value)
print('The average rainfall for the year:', avg)
```

This code is merely a combination of tools and functions that we have already introduced above and in earlier chapters, so nothing is truly new. Note, however, how we skip the first line with a single call to `infile.readline()`, which will simply read the first line and move to the next one, thus being ready to read the lines in which we are interested. If there are multiple header lines in the file we can simply add multiple `readline` calls to skip whatever we don't want to process. Notice also how list slicing is used to remove the yearly data from the lists. Negative indices in Python lists run backward, starting from the last element, so `annual_avg = rainfall[-1]` will extract the last value in the `rainfall` list and assign it to `annual_avg`. The list slicing `months[:-1]`, `rainfall[:-1]` will extract all elements from the lists up to, but not including the last one, thereby removing the yearly data from both lists.

## 5.4 Writing Data to Files

Writing data to files follows the same pattern as reading. We open a file for writing and typically use a for loop to traverse the data, which we then write to the file using `write`:

```
outfile = open(filename, 'w')   # 'w' for writing

for data in somelist:
    outfile.write(sometext + '\n')

outfile.close()
```

Notice the inclusion of \n in the call to `write`. Unlike `print`, a call to `write` will not by default add a line break after each call by defauls, so if we do not add this explicitly, the resulting file will consist of a single long line. It is often more convenient to have a line-structured file, and for this we include the \n, which adds a line break. The alternative way of opening files can also be used for writing, and it ensures that the file is automatically closed:

```
with open(filename, 'w') as outfile:   # 'w' for writing
    for data in somelist:
        outfile.write(sometext + '\n')
```

One should use caution when writing to files from Python programs. If you call `open(filename,'w')` with a filename that does not exist, a new file will be created; however, if a file with that name exists, it will simply be deleted and replaced by an empty file. Therefore, even if we do not actually write any data to the file, simply opening it for reading will erase all its contents. A safer way to write to files is to use 'open(filename,'a'), which will *append* data to the end of the file if it already exists, and create a new file if it does not exist.

For a concrete example, consider the task of writing information from a nested list to a file. We have following the nested list (rows and columns):

```
data = \
[[ 0.75,         0.29619813, -0.29619813, -0.75       ],
 [ 0.29619813,  0.11697778, -0.11697778, -0.29619813],
 [-0.29619813, -0.11697778,  0.11697778,  0.29619813],
 [-0.75,        -0.29619813,  0.29619813,  0.75       ]]
```

To write these data to a file in tabular form, we follow the steps outlined above and use a nested for loop (one for loop inside another) to traverse the list and write the data. The following code will do the trick:

```
with open('tmp_table.dat', 'w') as outfile:
    for row in data:
        for column in row:
            outfile.write(f'{column:14.8f}')
        outfile.write('\n')
```

The resulting file looks like

```
    0.75000000      0.29619813     -0.29619813     -0.75000000
    0.29619813      0.11697778     -0.11697778     -0.29619813
   -0.29619813     -0.11697778      0.11697778      0.29619813
   -0.75000000     -0.29619813      0.29619813      0.75000000
```

The nicely aligned columns are caused by the format specifier given to the f-string in the **write** call. The code will work fine without the format specifier, but the columns will not be aligned, and we also need to add a space after every number or, otherwise, each line will just be a long string of numbers that are difficult to separate. The structure of the nested for loop is also worth stepping through in the code above. The innermost loop traverses each row, writing the numbers one by one to the file. When this inner loop is done the program moves to the next line (**outfile.write('\n')**), which writes a linebreak to the file to end the line. After this line, one pass of the outer for loop is finished and the program moves to the next iteration and the next line in the table. The code for writing each number belongs inside the innermost loop, whereas the code for writing the line break is in the outer loop, since we only want one line break for each line.

## 5.5 Handling Errors in Programs

As demonstrated above, allowing user input in our programs will often introduce errors, and, as our programs grow in complexity, there can be multiple other sources of errors as well. Python has a general set of tools for handling such errors that is commonly referred to as *exception handling*, and it used in many different programming languages. To illustrate how it works, let us return to the example with the atmospheric pressure formula:

```
import sys
from math import exp

h = sys.argv[1]
h = float(h)

p0 = 100.0; h0 = 8400
print(p0 * exp(-h/h0))
```

As mentioned above, this code can easily break if the user provides a command line argument that cannot be converted to a float, that is, any argument that is not a pure number. Potentially even worse is our program failing with a fairly cryptic error message if the user does not include a command line argument at all, as in the following:

---
Terminal

```
Terminal> python altitude_cml.py
Traceback (most recent call last):
  File "altitude_cml.py", line 4, in ?
    h = sys.argv[1]
IndexError: list index out of range
```
---

How can we fix such problems and make the program more robust with respect to user errors? One possible solution is to add an if-test to check if any command line arguments have been included:

```
import sys
if len(sys.argv) < 2:
    print('You failed to provide a command line arg.!')
    exit()  # abort

h = float(sys.argv[1])

p0 = 100.0; h0 = 8400
print(p0 * exp(-h/h0))
```

The function call `exit()` will simply abort the program, so this extension solves part of the problem. The program will still stop if it is used incorrectly, but it will provide a more sensible and useful error message:

---
Terminal

```
Terminal> python altitude_cml.py
You failed to provide a command line arg.!
```
---

However, we only handle one of the potential errors, and using if-tests to test
for every possible error can lead to quite complex programs. Instead, it is
common in Python and many other languages to *try* to do what we intend
to and, if it fails, to recover from the error. This principle uses the try-except
block, which has the following general structure:

```
try:
    <statements we intend to do>
except:
    <statements for handling errors>
```

If something goes wrong in the `try` block, Python will raise an *exception* and
the execution jumps to the `except` block. Inside the `except` block, we need
to add our own code for *catching* the exception, basically to detect what went
wrong and try to fix it. If no errors occur inside the `try` block, the code inside
the `except` block is not run and the program simply moves on to the first
line after the try-except block.

**Improving the atmospheric pressure program with try-except.** To
apply the try-except idea to the air pressure program, we can try to read h
from the command line and convert it to a float, and, if this fails, we tell the
user what went wrong and stop the program:

```
import sys
try:
    h = float(sys.argv[1])
except:
    print('You failed to provide a command line arg.!')
    exit()

p0 = 100.0; h0 = 8400
print(p0 * exp(-h/h0))
```

One could argue that this is not very different from the program using the
if-test, but we shall see that the try-except block has some benefits. First, we
can try to run this program with different input, which immediately reveals
a problem:

---
Terminal
---

```
Terminal> python altitude_cml_except1.py
You failed to provide a command line arg.!

Terminal> python altitude_cml_except1.py 2469m
You failed to provide a command line arg.!
```

---

Regardless of what goes wrong inside our try block, Python will raise an
exception that needs to be handled by the except block. The problem with
our code is that all possible errors will be handled the same way. In the first
case, the problem is that there are no arguments, that is, `sys.argv[1]` does
not exist, which leads to an `IndexError`. This situation is correctly handled

by our code. In the second case, we provide an argument, so the indexing of `sys.argv` goes well, but the conversion fails, since Python does not know how to convert the string 2469m to a float. This is a different type of error, known as a `ValueError`, and we see that it is not treated very well by our except block. We can improve the code by letting the except block test for different types of errors, and handling each one differently:

```
import sys
try:
    h = float(sys.argv[1])
except IndexError:
    print('No command line argument for h!')
    sys.exit(1)  # abort execution
except ValueError:
    print(f'h must be a pure number, not {sys.argv[1]}')
    exit()

p0 = 100.0; h0 = 8400
print(p0 * exp(-h/h0))
```

The following two examples illustrate how this more specific error handling works:

---
Terminal
---

```
Terminal> python altitude.py
No command line argument for h!


Terminal> python altitude.py 2469m
The altitude must be a pure number, not "2469m"
```

---

Of course, a drawback of this approach is that we need to guess in advance what could go wrong inside the try-block, and write code to handle all possible errors. However, with some experience, this is usually not very difficult. Python has many built-in error types, but only a few that are likely to occur and which need to be considered in the programs we encounter throughout this book. In the code above, if the try block would leads to a different exception than what we catch in our except block, the code will simply end with a standard Python error message. If we want to avoid this behavior, and catch all possible exceptions, we could add a generic except block such as

```
except:
    print('Something went wrong in reading input data!')
    exit()
```

Such a block should be added after the `except ValueError` block in the code above, and will catch any exception that is not an `IndexError` nor a `ValueError`. In this particular case, it can be difficult to imagine what kind of error that would be, but if it occurs, it will be caught and handled by our generic except block.

**The programmer can also raise exceptions.** In the code above, the exceptions were raised by standard Python functions, and we wrote the code to catch them. Instead of just letting Python raise exceptions, we can raise our own and tailor the error messages to the problem at hand. We provide two examples of such use:

- Catching an exception, but raising a new one (re-raising) with an improved (tailored) error message.
- Raising an exception because of input data that we know are wrong, although Python accepts the data.

The basic syntax both for raising and re-raising an exception is `raise ExceptionType(message)`. The following code includes both examples:

```python
import sys

def read_altitude():
    try:
        h = float(sys.argv[1])
    except IndexError:
        # re-raise, but with specific explanation:
        raise IndexError(
           'The altitude must be supplied on the command line.')
    except ValueError:
        # re-raise, but with specific explanation:
        raise ValueError(
           f'Altitude must be number, not "{sys.argv[1]}".')

    # h is read correctly as a number, but has a wrong value:
    if h < -430 or h > 13000:
        raise ValueError(f'The formula is not valid for h={h}')
    return h
```

Here we have defined a function to handle the user input, but the code is otherwise quite similar to the previous examples. As above, the except blocks will catch two different types of error, but, instead of handling them (i.e., stopping the program), the blocks here will equip the exceptions with more specific error messages, and then pass them on to be handled somewhere else in our program. For this particular case, the difference is not very large, and one could argue that our first approach is simpler and therefore better; however, in larger programs it can often be better to re-raise exceptions and handle them elsewhere. The last part of the function is different, since the error raised here is not an error as far as Python is concerned. We can input any value of h into our formula, and, unless we input a large negative number, it will not give rise to a Python error[1]. However, as an estimate of air pressure the formula is only valid in the troposphere, the lower part of the Earth's atmosphere, which extends from the lowest point on Earth (on land), at 430

---

[1] If we set h to be a large negative number, the argument for the exp function becomes large and positive, and leads to an `OverflowError`. However, this error will occur only for values far outside the range of validity for our air pressure estimate.

m below sea level, to around 13 km above sea level. We can therefore let the program raise a `ValueError` for any h outside this range, even if it does not involve a Python error in the usual sense.

The following code shows how we can use the function above, and how we can catch and print the error message provided with the exceptions. The construction `except <error> as e` is used to access the error and use it inside the except block, as follows:

```
try:
    h = read_altitude()
except (IndexError, ValueError) as e:
    # print exception message and stop the program
    print(e)
    exit()
```

We can run the code in the terminal to confirm that we obtain the correct error messages:

---
Terminal
---
```
Terminal> python altitude_cml_except2.py
The altitude must be supplied on the command line.

Terminal> python altitude_cml_except2.py 1000m
Altitude must be number, not 1000m.

Terminal> python altitude_cml_except2.py 20000
The formula is not valid for h=20000.

Terminal> python altitude_cml_except2.py 8848
34.8773231887747
```
---

# 5.6 Making Modules

So far in this course we have frequently used modules such as `math` and `sys`, by importing them into our code:

```
from math import log
r = log(6)    # call log function in math module

import sys
x = eval(sys.argv[1])   # access list argv in sys module
```

Modules are extremely useful in Python programs, since they contain a collection of useful data and functions (as well as classes later), that we can reuse in our code. But what if you have written some general and useful functions yourself that you would like to reuse in more than one program? In such cases it would be convenient to make your own module that you can import into other programs when needed. Fortunately, this task is very simple in Python; just collect the functions you want in a file, and you have a new module!

To look at a specific example, say we want create a module containing the interest formula considered earlier and a few other useful formulas for computing with interest rates. We have the mathematical formulas

$$A = P(1+r/100)^n, \tag{5.1}$$

$$P = A(1+r/100)^{-n}, \tag{5.2}$$

$$n = \frac{\ln \frac{A}{P}}{\ln(1+r/100)}, \tag{5.3}$$

$$r = 100 \left( \left( \frac{A}{P} \right)^{1/n} - 1 \right), \tag{5.4}$$

where, as above, $P$ is the initial amount, $r$ is the interest rate (percent), $n$ is the number of years, and $A$ is the final amount. We now want to implement these formulas as Python functions and make a module of them. We write the functions in the usual way:

```python
from math import log as ln

def present_amount(P, r, n):
    return P*(1 + r/100)**n

def initial_amount(A, r, n):
    return A*(1 + r/100)**(-n)

def years(P, A, r):
    return ln(A/P)/ln(1 + r/100)

def annual_rate(P, A, n):
    return 100*((A/P)**(1.0/n) - 1)
```

If we now save these functions in a file `interest.py`, it becomes a module that we can import, just as we are used to with built-in Python modules. As an example, say we want to know how long it takes to double our money with an interest rate of 5%. The `years` function in the module provides the right formula, and we can import and use it in our program, as follows:

```python
from interest import years
P = 1; r = 5
n = years(P, 2*P, p)
print(f'Money has doubled after {n} years')
```

**We can add a *test block* to a module file.** If we try to run the module file above with `python interest.py` from the terminal, no output is produced since the functions are never called. Sometimes it can be useful to be able to add some examples of use in a module file, to demonstrate how the functions are called and used and give sensible output if we run the file with `python interest.py`. However, if we add regular function calls, print statements and other code to the file, this code will also be run whenever we import the module, which is usually not what we want. The solution is to add such example code in a *test block* at the end of the module file. The test block includes an if-test to check if the file is imported as a module or if it is run as a regular Python program. The code inside the test block is then executed only when the file is run as a program, and not when it is imported as a module into another program. The structure of the if-test and the test block is as follows:

```
if __name__ == '__main__': # this test defines the test block
    <block of statements>
```

The key is the first line, which checks the value of the built-in variable `__name__`. This string variable is automatically created and is always defined when Python runs. (Try putting `print(__name__)` inside one of your programs or type it in an interactive session.) Inside an imported module, `__name__` holds the name of the module, whereas in the main program its value is "`__main__`".

For our specific case, the complete test block can look like

```
if __name__ == '__main__':
    A = 2.31525
    P = 2.0
    r = 5
    n = 3
    A_ = present_amount(P, r, n)
    P_ = initial_amount(A, r, n)
    n_ = years(P, A, r)
    r_ = annual_rate(P, A, n)
    print(f'A={A_} ({A}) P={P_} ({A}) n={n_} ({n}) r={r_} ({p})')
```

Test blocks are often included simply for demonstrating and documenting how modules are used, or they are included in files that we sometimes use as stand-alone programs and sometimes as modules. As indicated by the name, they are also frequently used to test modules. Using what we learned about test functions in the previous chapter, we can do this by writing a standard test function that tests the functions in the module, and then simply calling this function from inside the test block:

```
def test_all_functions():
    # Define compatible values
    A = 2.31525; P = 2.0; r = 5.0; n = 3
    # Given three of these, compute the remaining one
    # and compare with the correct value (in parenthesis)
    A_computed = present_amount(P, r, n)
    P_computed = initial_amount(A, r, n)
    n_computed = years(P, A, r)
    r_computed = annual_rate(P, A, n)
    def float_eq(a, b, tolerance=1E-12):
```

```
        """Return True if a == b within the tolerance."""
        return abs(a - b) < tolerance

    success = float_eq(A_computed,  A)  and \
              float_eq(A0_computed, A0) and \
              float_eq(p_computed,  p)  and \
              float_eq(n_computed,  n)
    assert success  # could add message here if desired

if __name__ == '__main__':
    test_all_functions()
```

Since we have followed the naming convention of test functions, the function will be called if we run, for instance, `pytest interest.py`, but since we call it from inside the test block, the test can also be run simply by `python interest.py`. In the latter case, the test will produce no output unless there are errors. However, if we import the module to use in another program, the test function is not run, because the variable `__name__` will be the name of the module (i.e. `interest`) and the test `__name__ == '__main__'` will be evaluated as false.

**How Python finds our new module.** Python has a number of designated places where it looks for modules. The first place it looks is in the same folder as the main program; therefore, if we put our module files there, they will always be found. However, this is not very convenient if we write more general modules that we plan to use from several other programs. Such modules can be put in a designated directory, say `/Users/sundnes/lib/python/mymods` or any other directory name that you choose. Then we need to tell Python to look for modules in this directory; otherwise, it will not find the module. On Unix-like systems (Linux, Mac, etc.), the standard way to tell Python where to look is by editing the *environment variable* called `PYTHONPATH`. Environment variables are variables that hold important information used by the operating system, and `PYTHONPATH` is used to specify the folders where Python should look for modules. If you type `echo $PYTHONPATH` in the terminal window, you will most likely obtain no output, since you have not added any folder names to this variable. We can put our new folder name in this variable by running the command

---
Terminal

```
export PYTHONPATH=/Users/sundnes/lib/python/mymods
```
---

However, if the `PYTHONPATH` already contained any folders, these will now be lost; therefore, to be on the safe side, it is better to use

---
Terminal

```
export PYTHONPATH=$PYTHONPATH:/Users/sundnes/lib/python/mymods
```
---

This last command will simply add our new folder to the end of what is already in our `PYTHONPATH` variable. To avoid having to run this command every time we want to import a module, we can put it in the file `.bashrc`, to ensure that it is run automatically when we open a new terminal window. The `.bashrc` file should be in your home directory (e.g. '/Users/sundnes/.bashrc'), and will be listed with `ls -a`. (The dot at the start of the filename makes it a *hidden* file, so it will not show up with just `ls`.) If the file is not there, you can simply create it in an editor and save it in your home directory, and the system should find it and read it automatically the next time you open a terminal window. As an alternative to editing the systemwide environment variable, we can also add our directory to the path from inside the program. Putting a line such as this inside your code, before you import the module, should allow Python to find it:

```
sys.path.insert(0, '/Users/sundnes/lib/python/mymods')
```

As an alternative to creating your own directory for modules, and then tell Python where to find them, you can place the modules in one of the places where Python always looks for modules. The location of these varies a bit between different Python installations, but the directory itself is usually named `site-packages`. If you have installed NumPy[2] or another package that is not part of the standard Python distribution, you can locate the correct directory by importing this package. For instance, type the following in an interactive Python shell:

```
>>> import numpy
>>> numpy.__file__
'/Users/sundnes/anaconda3/lib/python3.7/site-packages/numpy/__init__.py'
>>>
```

The last line reveals the location of the `site-packages` directory, and placing your own modules there will ensure Python will find them.

---

[2]NumPy is a package for numerical calculations. It is not part of the standard Python distribution, but it is often installed automatically if you install Python from other sources, for instance, from Anaconda. Otherwise, it can be installed for instance, with `pip` or other tools. The NumPy package will be used extensively in the next chapter.

# Chapter 6
# Arrays and Plotting

In this chapter, we will learn to visualize mathematical functions and the results of mathematical calculations. You have probably used a variety of different plotting tools in the past, and we will now do much of the same thing in Python. The way standard plotting tools work in Python is that we first compute a number of points lying on the curve we want to show and then draw straight lines between them. If we have enough points, the result looks like a smooth curve. For plotting mathematical functions, this approach can seem a bit primitive, since there are other tools we can use to simply type in a mathematical expression and have the curve plotted on the screen. However, the approach we use here is also much more flexible, since we can plot data when there is no underlying mathematical function, for instance, experimental data read from a file or results from a numerical experiment. To plot functions in Python, we need to learn about the package `matplotlib`, which is an extensive toolbox for plotting and visualization. In addition, it is useful to introduce the package named NumPy, which is useful for storing storing *arrays* of data for efficient computations.

## 6.1 NumPy and Array Computing

The standard way to plot a curve $y = f(x)$ is to draw straight lines between points along the curve, and for this purpose we need to store the coordinates of the points. We could use lists for this, for instance, two lists x and y, and most of the plotting tools we will use work fine with lists. However, a data structure known as an *array* is much more efficient than a list, and it offers a number of nice features and advantages. Computing with arrays is often referred to as *array computations* or *vectorized computations*, and these concepts are useful for much more than just plotting curves.

**Arrays are generalizations of vectors.** In high school mathematics, vectors were introduced as line segments with a direction, represented by coor-

J. Sundnes, *Introduction to Scientific Programming with Python*, Simula SpringerBriefs on Computing 6, https://doi.org/10.1007/978-3-030-50356-7_6

dinates $(x, y)$ in the plane or $(x, y, z)$ in space. This concept of vectors can be generalized to any number of dimensions, and we can view a vector $v$ as a general $n$-tuple of numbers; $v = (v_0, \ldots, v_{n-1})$. In Python, we could use a list to represent such a vector, by storing component $v_i$ as element `v[i]` in the list. However, vectors are so useful and common in scientific programming that a special data structure has been created for them: the *NumPy array*. An array is much less flexible than a list, in that it has a fixed length (i.e., no `append`-method), and one array can only hold variables of the same type. However, arrays are also much more efficient to use in computations, and since they are designed for such use, they have a number of useful features that can shorten and clarify our code.

For the purpose of plotting, we will mostly use one-dimensional arrays, but an array can have multiple indices, similar to a nested list. For instance, a two-dimensional array $A_{i,j}$ can be viewed as a table of numbers, with one index for the row and one for the column, as follows:

$$\begin{bmatrix} 0 & 7 & -3 & 5 \\ -1 & -3 & 4 & 0 \\ 9 & 3 & 5 & -7 \end{bmatrix} \qquad A = \begin{bmatrix} A_{0,0} & \cdots & A_{0,n-1} \\ \vdots & \ddots & \vdots \\ A_{m-1,0} & \cdots & A_{m-1,n-1} \end{bmatrix}$$

Such a two-dimensional case is similar to a matrix in linear algebra, but NumPy arrays do not follow the standard rules for mathematical operations on matrices. The number of indices in an array is often referred to as the *rank* or the *number of dimensions*.

**Storing (x,y) points on a curve in lists and arrays.** To make the array concept a bit more concrete, we consider the task mentioned above, where we want to store points on a function curve $y = f(x)$. All the plotting cases we will consider are based on this idea, so it makes sense to introduce it for a simple example. We have seen in previous chapters that there are multiple ways to store such pairs of numbers, for instance in a nested list containing $(x, y)$ pairs. However, for the purpose of plotting, the easiest approach is to create two lists or arrays, one holding the $x$-values and another holding the $y$-values. The two lists/arrays should be of equal length, and we will always create them using the same two steps. First, we create $n$ uniformly spaced $x$-values that cover the interval where we want to plot the function. Then, we run through these numbers and compute the corresponding $y$-values, storing these in a separate list or array. The following interactive session illustrates the procedure, using list comprehensions to first create a list of five $x$-points on the interval $[0, 1]$, and then compute the corresponding points $y = f(x)$ for $f(x) = x^2$.

```
>>> def f(x):
...     return x**2
...
>>> n = 5                    # number of points
>>> dx = 1.0/(n-1)           # x spacing in [0,1]
```

```
>>> xlist = [i*dx for i in range(n)]
>>> ylist = [f(x) for x in xlist]
```

Now that we have the two lists, they can be sent directly to a tool such as
`matplotlib` for plotting, but before we do this, we will introduce NumPy
arrays. If we continue the interactive session from above, the following lines
will turn the two lists into NumPy arrays:

```
>>> import numpy as np        # module for arrays
>>> x = np.array(xlist)       # turn list xlist into array
>>> y = np.array(ylist)
```

It is worth noting how we import NumPy in the first line. As always, we could
import it with `from numpy import *`, but this is a bad habit, since `numpy`
and `math` contain many functions with the same name, and we will often use
both modules in the same program. To ensure that we always know which
module we are using, it is a good habit to import NumPy as we have done
here. Using `import numpy as np` instead of simply `import numpy` saves us
some typing in the rest of the code and is also more or less an accepted
standard among Python programmers.

Converting lists to arrays using the `array` function from NumPy is intuitive and flexible, but NumPy has a number of built-in functions that are
often more convenient to use. Two of the most widely used ones are called
`linspace` and `zeros`. The following interactive session is a list-free version of
the example above, where we create the NumPy arrays directly, using these
two functions:

```
>>> import numpy as np
>>> def f(x):
...     return x**2
...
>>> n = 5                     # number of points
>>> x = np.linspace(0, 1, n)  # n points in [0, 1]
>>> y = np.zeros(n)           # n zeros (float data type)
>>> for i in range(n):
...     y[i] = f(x[i])
...
```

As illustrated here, we will usually call `linspace` with three arguments, with
the general form `linspace(start,stop,n)`, which will create an array of
length n, containing uniformly distributed values on the interval from `start`
to `stop`. If we leave out the third argument, as in `linspace(start,stop)`,
a default value of n=50 is used. The `start` and `stop` arguments must always
be provided. An array of equally spaced $x$-values is needed nearly every time
we plot something, so we will use `linspace` frequently. It is worth spending
time to become familiar with how it is used and what it returns.

The second NumPy function used above, `zeros(n)`, does exactly what we
would expect: it creates an array of length n containing only zeros. We have
seen earlier that a common way to create a list is to start with an empty
list and fill it with values using a for loop and the `append`-method. We will

often use a similar approach to create an array, but since an array has fixed length and no `append`-method, we must first create an array of the right size and then loop over it with an index to fill in the values. This operation is very common, so remembering the existence of NumPy's `zeros` function is important.

As we have seen in Chapter 3, lists in Python are extremely flexible, and can contain any Python object. Arrays are much more static, and we will typically use them for numbers (i.e., type `float` or `int`). They can also be of other types, such as boolean arrays (true/false), but a single array always contains a single object type. We have also seen that arrays are of fixed length and do not have the convenient `append`-method. So, why do we use arrays at all? One reason, which was mentioned above, is that arrays are more efficient to store in memory and use in computations. The other reason is that arrays can shortn our code and make it more readable, since we can perform operations on an entire array at once instead of using loops. Say, for instance, that we want to compute the cosine of all the elements in a list or array `x`. We know how to do this using a for loop

```
import numpy as np
from math import cos
x = np.linspace(0,1,11)

for i in range(len(x)):
    y[i] = cos(x[i])
```

but if `x` is an array, `y` can be computed by

```
y = np.cos(x)                    # x: array, y: array
```

In addition to being shorter and quicker to write, this code will run much faster than the code with the loop.[1] Such computations are usually referred to as vectorized computations, since they work on the entire array (or vector) at once. Most of the standard functions we find in `math` have a corresponding function in `numpy` that will work for arrays. Under the hood these NumPy functions still contain a for loop, since they need to traverse all the elements of the array, but this loop is written in very efficient C code and is therefore much faster than Python loops we write ourselves.

A function `f(x)` that was written to work a for a single number `x` will often work well for an array as well. If the function uses only basic mathematical operators $(+, -, *,$ etc.), we can pass it either a number or an array as the argument, and it will work just fine with no modifications. If the function uses more advanced operations that we need to import, we have to make sure to

---

[1] For the small array considered here, containing just 11 numbers, the efficiency gain does not matter at all. It will be difficult to detect a difference between the two versions even if we measure the run time of our program. However, certain numerical programs can use nested arrays containing tens of millions of numbers, and in such cases the difference between loops and vectorized code becomes very noticeable.

import these from `numpy` rather than `math`, since the functions in `math` work only with single numbers. The following example illustrates how it works:

```
from numpy import sin, exp, linspace

def g(x):
    return x**2+2*x-4

def f(x):
    return sin(x)*exp(-2*x)

x = 1.2                      # float object
y = f(x)                     # y is float

x = linspace(0, 3, 101)      # 100 intervals in [0,3]
y = f(x)                     # y is array
z = g(x)             # z is array
```

We see that, except for the initial import from NumPy, the two functions look exactly the same as if they were written to work on a single number. The result of the two function calls will be two arrays `y,z` of length 101, with each element being the function value computed for the corresponding value of `x`.

If we try to send an array of length $> 1$ to a function imported from `math`, we will obtain an error message:

```
>>> import math, numpy
>>> x = numpy.linspace(0, 1, 6)
>>> x
array([0. , 0.2, 0.4, 0.6, 0.8, 1. ])
>>> math.cos(x[0])
1.0
>>> math.cos(x)
Traceback (most recent call last):
  File "<stdin>", line 1, in <module>
TypeError: only size-1 arrays can be converted to Python scalars
>>> numpy.cos(x)
array([1.        , 0.98006658, 0.92106099, 0.82533561, 0.69670671,
       0.54030231])
```

On the other hand, using NumPy functions on single numbers will work just fine. A natural question to ask, then, is why do we ever need to import from `math` at all? Why not use NumPy functions all the time, since they do the job for both arrays and numbers? The answer is that we can certainly do this, and in most cases it will work fine, but the functions in `math` are more optimized for single numbers (scalars) and are therefore faster. One will rarely notice the difference, but there can be applications where the extra efficiency matters. There are also functions in `math` (e.g., `factorial`) that do not have a corresponding version in NumPy.

We started this chapter by computing points along a curve using lists and for loops. Now that we have introduced NumPy, we can solve this task much more easily by using arrays and array computations. Say we want to compute points on the curve described by the function

$$f(x) = e^{-x}\sin(2\pi x), \quad x \in [0,4]$$

for $x \in [0, 4*\pi]$. The vectorized code can look as follows:

```
import numpy as np

n = 100
x = np.linspace(0, 4, n+1)
y = np.exp(-x)*np.sin(2*np.pi*x)
```

This code is shorter and quicker to write than the one with lists and loops, most people find it easier to read since it is closer to the mathematics, and it runs much faster than the list version.

We have already mentioned the term *vectorized computations*, and if you follow a course in scientific Python you will probably be asked at some point to *vectorize* a function or a computation. This usually means nothing more than to ensure that all the mathematical functions are imported from numpy rather than math, and to then perform all the operations on entire arrays rather than looping over their individual elements. The vectorized code should contain no for loops written in Python. The mathematical functions g(x) and f(x) in the example above are perfectly valid examples of vectorized functions, even though the actual functions look identical to the scalar versions. The only major exceptions to this simple recipe for vectorization are functions that include if-tests. For instance, in Chapter 4, we implemented piecewise-defined mathematical functions using if-tests. These functions will not work if the input argument is an array, because a test such as if x > 0 has no precise meaning if x is an array. There are ways, however, to solve this problem, which we will look into later in the chapter.

## 6.2 Plotting Curves with Matplotlib

The motivation for introducing NumPy arrays was to plot mathematical functions, and now that we have introduced all the necessary tools we are finally ready to do so. Let us start with a simple example. Say we want to plot the curve $y(x) = e^{-x}\sin(2\pi x)$, for $x$ ranging from zero to four. The code can look like

```
import matplotlib.pyplot as plt
import numpy as np

n = 100
x = np.linspace(0, 4, n+1)
y = np.exp(-x)*np.sin(2*np.pi*x)
```

```
plt.plot(x, y)
plt.show()
```

This code is identical to the example above, except for the first line and the last two lines. The first line imports the plotting tools from the `matplotlib` package, which is an extensive library of functions for scientific visualization. We will only use a small subset of the capabilities of `matplotlib`, mostly from the module `pyplot`, to plot curves and create animations of curves that change over time. The next few lines are from the example above, and they simply create the two arrays `x` and `y` defining the points along the curve. The last two lines carry out the actual plotting: the call `plt.plot(x,y)` first creates the plot of the curve, and then `plt.show()` displays the plot on the screen. The reason for keeping these separate is to make it easy to plot multiple curves in a single plot, by calling `plot` multiple times followed by a single call to `show`. The resulting plot is shown in Figure 6.1. A common mistake is to forget the `plt.show()` call, and the program will then simply end without displaying anything on the screen.

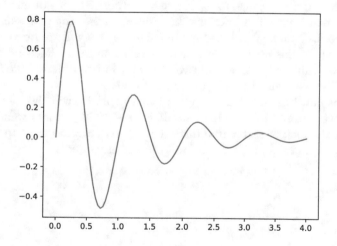

**Fig. 6.1** Simple plot of a function using Matplotlib.

The plot produced by the code above is very simple and contains no title, axis labels, or other information. We can easily add such information in the plot by using tools from `matplotlib`:

```
import matplotlib.pyplot as plt  # import and plotting
import numpy as np

def f(x):
```

```
      return np.exp(-x)*np.sin(2*np.pi*x)

n = 100
x = np.linspace(0, 4, n+1)
y = f(x)

plt.plot(x, y, label='exp(-x)*sin(2$\pi$ x)')

plt.xlabel('x')                # label on the x axis
plt.ylabel('y')                # label on the y axis
plt.legend()                   # mark the curve
plt.axis([0, 4, -0.5, 0.8])    # [tmin, tmax, ymin, ymax]
plt.title('My First Matplotlib Demo')

plt.savefig('fig.pdf')     # make PDF image for reports
plt.savefig('fig.png')     # make PNG image for web pages
plt.show()
```

The plot resulting from this code is shown in Figure 6.2. Most of the lines
in the code should be self-explanatory, but some are worth a comment. The
call to `legend` will create a legend for the plot, using the information provided in the `label` argument passed to `plt.plot`. This is very useful when
plotting multiple curves in a single plot. The `axis` function sets the length
of the horizontal and vertical axes. These are otherwise set automatically by
Matplotlib, which usually works fine, but in some cases the plot looks better
if we set the axes manually. Later in this chapter, we will create animations
of curves and, in this case, the axes will have to be set to fixed lengths. Finally, the two calls to `savefig` will save our plot in two different file formats,
automatically determined by the file name provided.

If we plot multiple curves in a single plot, Matplotlib will choose the color
of each curve automatically. This default choice usually works well, but we
can control the look of each curve further if desired. Say we want to plot the
functions $e^{-x}\sin(2\pi x)$ and $e^{-2x}\sin(4\pi x)$ in the same plot:

```
import matplotlib.pyplot as plt
import numpy as np

def f1(x):
    return np.exp(-x)*np.sin(2*np.pi*x)

def f2(x):
    return np.exp(-2*x)*np.sin(4*np.pi*x)

x = np.linspace(0, 8, 401)
y1 = f1(x)
y2 = f2(x)

plt.plot(x, y1, 'r--', label='exp(-x)*sin(2$\pi$ x)')
plt.plot(x, y2, 'g:', label='exp(-2*x)*sin(4$\pi$ x)')

plt.xlabel('x')
plt.ylabel('y')
```

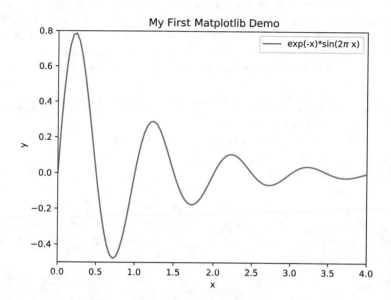

**Fig. 6.2** Example plot with more information added.

```
plt.legend()
plt.title('Plotting two curves in the same plot')
plt.savefig('fig_two_curves.png')
plt.show()
```

This example shows that the options for changing the color and plotting style of the curves are fairly intuitive, and can be easily explored by trial and error. For a full overview of all the options, we refer the reader to the Matplotlib documentation.

Although the code example above was not too complex, we had to write an excess of 20 lines just to plot two simple functions on the screen. This level of programming is necessary if we want to produce professional-looking plots, such as in a presentation, master's thesis, or scientific report. However, if we just want a quick plot on the screen, this can be achieved much more simply. The following code lines will plot the same two curves as in the example above, using just three lines:

```
t = np.linspace(0, 8, 201)
plt.plot(x,np.exp(-x)*np.sin(2*np.pi*x),x,np.exp(-2*x)*np.sin(4*np.pi*x))
plt.show()
```

As always, the effort we put in depends on what the resulting plot will be used for, and, in particular, on whether we are just exploring some data on our own or plan on presenting it to others.

**Example: Plotting a user-specified function.** Say we want to write a small program `plotf.py` that asks the user to provide a mathematical function $f(x)$, and then plots the curve $y = f(x)$. We can also ask the user to specify the boundaries of the curve, that is, the lower and upper limits for $x$. An example of running the program from the terminal can look like should be

```
Terminal
Terminal> python plot_input.py
Write a mathematical expression of x:2*x**2-4
Provide lower bound for x:0
Provide upper bound for x:7
```

For these input values the program should plot the curve $y = 2x^2 - 4$, for $x \in [0,7]$. The `plot_input.py` program should work for any mathematical expression. The task can be solved using the the functions `input` and `eval` introduced in Chapter 5:

```python
from numpy import *
import matplotlib.pyplot as plt

formula = input('Write a mathematical expression of x:')
xmin = float(input('Provide lower bound for x:'))
xmax = float(input('Provide upper bound for x:'))

x = linspace(xmin, xmax, 101)
y = eval(formula)

plt.plot(x, y)
plt.show()
```

This small program will stop and ask the user first for a mathematical expression and then for the bounds on `x`, and then it will proceed to plot the resulting curve. Note that, in this case, we have a good reason to import NumPy with `from numpy import *`. We want the user to be able type a formula using standard mathematical terminology, such as `sin(x) + x**2` (rather than `np.sin(x) + x**2`). For this to work, we need to import all the mathematical functions from NumPy without a prefix.

# 6.3 Plotting Discontinuous and Piecewise-Defined Functions

Discontinuous functions, and functions defined in a piecewise manner, are common in science and engineering. We saw in Chapter 4 how these could be implemented in Python using if-tests, but, as we briefly commented above, this implementation gives rise to challenges when using arrays and NumPy.

To consider a concrete example, say we want to plot the Heaviside function, defined by

$$H(x) = \begin{cases} 0, \; x < 0 \\ 1, \; x \geq 0 \end{cases}$$

Following the ideas from Chapter 4, a Python implementation of this function could look like this

```
def H(x):
    if x < 0:
        return 0
    else:
        return 1
```

Now we want to plot the function using the simple approach introduced above. It is natural to simply create an array of values x, and to pass this array to the function H(x) to compute the corresponding $y$-values:

```
x = linspace(-10, 10, 5)   # few points (simple curve)
y = H(x)
plot(x, y)
```

However, if we try to run this code, we obtain an error message, a `ValueError` error inside the function II(x), coming from the `if x < 0` line. We can illustrate what goes wrong in an interactive Python session:

```
>>> x = linspace(-10,10,5)
>>> x
array([-10.,   5.,   0.,   5.,  10.])
>>> b = x < 0
>>> b
array([ True,  True, False, False, False], dtype=bool)
>>> bool(b)   # evaluate b in a Boolean context
...
ValueError: The truth value of an array with more than
one element is ambiguous. Use a.any() or a.all()
```

We see here that the result of the statement b = x < 0 is an array of Boolean values, whereas, if b were a single number, the result would be a single Boolean (true/false). Therefore, the statement `bool(b)`, or tests such as if b or if x < 0 do not make sense, since it is impossible to say whether an array of multiple true/false values is true or false.

There are several ways to fix this problem. One is to avoid the vectorization altogether, and return to the traditional for loop for computing the values:

```
import numpy as np
import matplotlib.pyplot as plt
n = 5
x = np.linspace(-5, 5, n+1)
y = np.zeros(n+1)

for i in range(len(x)):
    y[i] = H(x[i])
```

```
plt.plot(x,y)
plt.show()
```

A variation of the same approach is to alter the H(x) function itself and put the for loop inside it:

```
def H_loop(x):
    r = np.zeros(len(x))   # or r = x.copy()
    for i in range(len(x)):
        r[i] = H(x[i])
    return r

n = 5
x = np.linspace(-5, 5, n+1)
y = H_loop(x)
```

We see that this last approach ensures that we can call the function with an array argument x, but the downside to both versions is that we need to write quite a lot of new code, and using a for loop is much slower than using vectorized array computing.

An alternative approach is to use a built-sin NumPy function named vectorize[2], which offers automatic vectorization of functions with if-tests. The line

```
Hv = np.vectorize(H)
```

creates a vectorized version Hv(x) of the function H(x) that will work with an array argument. Although this approach is obviously better, in the sense that the conversion is automatic so we need to write very little new code, it is about as slow as the two approaches using for loops.

A third approach is to write a new function where the if-test is coded differently:

```
def Hv(x):
    return np.where(x < 0, 0.0, 1.0)
```

For this particular case, the NumPy function where will evaluate the expression x<0 for all elements in the array x, and return an array of the same length as x, with values 0.0 for all elements where x<0, and 1.0 for the others. More generally, a function with an if-test can be converted to an array-ready vectorized version in the following way:

```
def f(x):
```

---

[2]It is a fairly common misconception to believe that vectorizing a computation or making a vectorized version of a function, always involves using the function numpy.vectorize. This is not the case. In most cases, we only need to make sure that we use array-ready functions, such as numpy.sin, numpy.exp, etc., instead of the scalar version from math, and code all calculations so that they work on an entire array instead of stepping through the elements with a for loop. The vectorize-function is usually only necessary for functions containing if-tests.

```
    if condition:
        x = <expression1>
    else:
        x = <expression2>
    return x

def f_vectorized(x):
    x1 = <expression1>
    x2 = <expression2>
    r = np.where(condition, x1, x2)
    return r
```

This conversion is not, of course, as automatic as using `vectorize`, and requires writing some more code, but it is much more computationally efficient than the other versions. Efficiency is sometimes important when working with large arrays.

## 6.4 Making a Movie of a Plot

It is often useful to make animations or movies of plots, for instance if the plot represents some physical phenomenon that changes with time, or if we want to visualize the effect of changing parameters. Matplotlib has multiple tools for creating such plots, and we will explore some of them here. To start with a specific case, consider again the well-known Gaussian bell function:

$$f(x; m, s) = \frac{1}{\sqrt{2\pi}} \frac{1}{s} \exp\left[ -\frac{1}{2} \left( \frac{x - m}{s} \right)^2 \right]$$

The parameter $m$ is the location of the function's peak, while $s$ is a measure of the width of the bell curve. Plots of this function for different values of $s$ are shown in Figure 6.3. As an alternative illustration of how the parameters change the function we can make a movie (animation) of how $f(x; m, s)$ changes shape as $s$ goes from two to 0.2.

**Movies are made from a large set of individual plots.** Movies of plots are created through the classical approach of cartoon movies (or, really, all movies): by creating a set of images and viewing them in rapid sequence. For our specific example, the typical approach is to write a for loop to step through the $s$ values and either show the resulting plots directly or store them in individual files for later processing. Regardless of the approach, it is important to always fix the axes when making animations of plots; otherwise, the $y$ axis always adapts to the peak of the function and the visual impression is completely wrong

We will look at three different ways to create a movie of the kind outlined above:

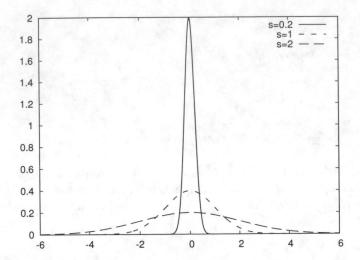

**Fig. 6.3** The Gaussian bell function plotted for different values of $s$.

1. Let the animation run *live*, without saving any files. With this approach, the plots are simply drawn on the screen as they are created, that is, one plot is shown for each pass of the for loop. The approach is simple, but has the disadvantage that we cannot pause the movie or change its speed.
2. Loop over all data values, create one plot for each value and save it to a file, and then combine all the image files into a movie. This approach enables us to actually create a movie file that can be played using standard movie player software. The drawback of this approach is that it requires separately installed software (e.g., *ImageMagick*) to create the movie and view the animation.
3. Use a `FuncAnimation` object from Matplotlib. This approach uses a slightly more advanced feature of Matplotlib, and can be considered a combination of the two approaches above. The animation is played *live*, but it can also be stored in a movie file. The downside is that the creation of the movie file still relies on externally installed software that needs to be installed separately and integrated with Matplotlib.

**First alternative: Running the movie live as the plots are created.**
This approach is the simplest of the three and requires very few tools that we have not already seen. We simply use a for loop to loop over the $s$ values, compute new $y$-values, and update the plot for each iteration of the loop. However, we need to be aware of a couple of technical details. In particular, the intuitive approach of simply including calls to `plot(x,y)` followed by `show()` inside the for loop does not work. Calling `show()` will make the program stop after the first plot is drawn, and it will not run further until we close the plotting window. Additionally, recall that we used multiple calls to

plot when we wanted multiple curves in a single window, which is not what
we want here. Instead, we need to create an object that represents the plot
and then update the $y$-values of this object for each pass through the loop.
The complete code can look like

```
import matplotlib.pyplot as plt
import numpy as np

def f(x, m, s):
    return (1.0/(np.sqrt(2*np.pi)*s))*np.exp(-0.5*((x-m)/s)**2)

m = 0;  s_start = 2;  s_stop = 0.2
s_values = np.linspace(s_start, s_stop, 30)

x = np.linspace(m -3*s_start, m + 3*s_start, 1000)
# f is max for x=m (smaller s gives larger max value)
max_f = f(m, m, s_stop)

y = f(x,m,s_stop)
lines = plt.plot(x,y)   #Returns a list of line objects!

plt.axis([x[0], x[-1], -0.1, max_f])
plt.xlabel('x')
plt.ylabel('f')

for s in s_values:
    y = f(x, m, s)
    lines[0].set_ydata(y) #update plot data and redraw
    plt.draw()
    plt.pause(0.1)
```

Most of the lines in this code should be familiar, but there are a few items
that are worth noting. First, we use the same plot function as earlier, but in
a slightly different manner. Generally, this function does two things: it creates
a plot that is ready to display on the screen by a subsequent call to show(),
and it returns a special Matplotlib object that represents the plot (a Line2D
object). In the examples above, we did not need this object, so we did not
care about it, but this time we store it in the variable lines. Note also that
the plot-function always returns a list of such objects, representing all the
curves of the plot. In this case, we plot only one curve, and the list has length
one. To update the plot inside the for loop, we call the set_ydata method of
this object, that is, lines[0].set_ydata(y), every time we have computed
a new y array. After updating the data, we call the function draw() to draw
the curve on the screen. The final line inside the for loop simply makes the
program stop and wait for 0.1 seconds. If we remove this call, the movie runs
too fast to be visible, and we can obviously adjust the speed by changing
the function's argument. As a final comment on this code, remember the
important message from above, that we always need to fix the axes when
creating movies; otherwise, Matplotlib will adjust the axes automatically for
each plot, and the resulting movie will not really look like a movie at all.

Here, we compute the maximum value that the function will obtain in the line `max_f = f(m, m, s_stop)` (based on either prior knowledge about the Gaussian function or inspection of the mathematical expression). This value is then used to set the axes for all the plots that make up the movie.

**Second alternative: Saving image files for later processing.** This approach is very similar to the one above, but, instead of showing the plots on the screen, we save them to files, using the `savefig` function from Matplotlib. To avoid having each new plot over-write the previous file, we must include a counter variable and a formatted string to create a unique filename for each iteration of the for loop. The complete code is nearly identical to the one above:

```python
import matplotlib.pyplot as plt
import numpy as np

def f(x, m, s):
    return (1.0/(np.sqrt(2*np.pi)*s))*np.exp(-0.5*((x-m)/s)**2)

m = 0;  s_start = 2;  s_stop = 0.2
s_values = np.linspace(s_start, s_stop, 30)

x = np.linspace(m -3*s_start, m + 3*s_start, 1000)
# f is max for x=m (smaller s gives larger max value)
max_f = f(m, m, s_stop)

y = f(x,m,s_stop)
lines = plt.plot(x,y)

plt.axis([x[0], x[-1], -0.1, max_f])
plt.xlabel('x')
plt.ylabel('f')

frame_counter = 0
for s in s_values:
    y = f(x, m, s)
    lines[0].set_ydata(y) #update plot data and redraw
    plt.draw()
    plt.savefig(f'tmp_{frame_counter:04d}.png') #unique filename
    frame_counter += 1
```

Running this program should create a number of image files, all located in the directory in which we run the program. Converting these images into a movie requires external software, for instance, `convert` from the ImageMagick software suite to make animated gifs, or `ffmpeg` or `avconv` to make MP4 and other movie formats. For instance, if we want to create an animated gif of the image files produced above, the following command will do the trick:

--------------------------------- Terminal ---------------------------------

Terminal> convert -delay 20 tmp_*.png movie.gif

----------------------------------------------------------------------------

The resulting gif can be played using `animate` from ImageMagick or in a browser. Note that, for this approach to work, one needs to be careful about the filenames. The argument `tmp_*.png` passed to the convert function will simply replace * with any text, thereby sending all files with this pattern to `convert`. The files are sent in lexicographic (i.e., alphabetical) order, which is why we use the format specifier `04d` in the f-string above. It would be tempting so simply write `{frame_counter}`, with no format specifier, inside the f-string to create the unique filename, and not worry about the format specifier. This approach would create unique filenames such as `tmp_0.png`, `tmp_1.png`, and so on. However, we would run into problems when creating the movie with `convert`, since, for instance, `tmp_10.png` comes before `tmp_9.png` in the alphabetic ordering.

**Third alternative: Using built-in Matplotlib tools.** The third approach is the most advanced and flexible, and it relies on built-in Matplotlib tools instead of the explicit for loop that we used above. Without an explicit for loop, the actual steps of creating the animation are less obvious, and the approach is therefore somewhat less intuitive. The essential steps are the following:

1. Make a function to update the plot. In our case, this function should compute the new y array and call `set_ydata`, as above, to update the plot.
2. Make a list or array of the argument that changes (in this case, $s$).
3. Pass the function and the list as arguments to create a `FuncAnimation` object.

After creating this object, we can use various built-in methods to save the movie to a file, show it on the screen, and so forth. The complete code looks like the following:

```
import numpy as np
import matplotlib.pyplot as plt
from matplotlib.animation import FuncAnimation

def f(x, m, s):
    return (1.0/(np.sqrt(2*np.pi)*s))*np.exp(-0.5*((x-m)/s)**2)

m = 0; s_start = 2; s_stop = 0.2
s_values = np.linspace(s_start,s_stop,30)

x = np.linspace(-3*s_start,3*s_start, 1000)

max_f = f(m,m,s_stop)

plt.axis([x[0],x[-1],0,max_f])
plt.xlabel('x')
plt.ylabel('y')

y = f(x,m,s_start)
```

```
lines = plt.plot(x,y) #initial plot to create the lines object

def next_frame(s):
    y = f(x, m, s)
    lines[0].set_ydata(y)
    return lines

ani = FuncAnimation(plt.gcf(), next_frame, frames=s_values, interval=100)
ani.save('movie.mp4',fps=20)
plt.show()
```

Most of the lines are identical to the examples above, but there are some key differences. We define a function `next_frame` that contains all the code that updates the plot for each frame, and returns an updated `Line2D` object. The argument to this function should be whatever argument that is changed for each frame (in our case, `s`). After defining this function, we use it to create a `FuncAnimation` object in the next line:

```
ani = FuncAnimation(plt.gcf(), next_frame, frames=s_values, interval=100)
```

This function call returns an object of type `FuncAnimation` [3]. The first argument is simply the current figure object we are working with (`gcf` being short for *get current figure*), the next is the function we just defined to update the frames, the third is the array of `s`-values used to create the plots, and the last argument is the interval between frames in milliseconds. Numerous other optional arguments to the function can be used to tune the animation. We refer to the Matplotlib documentation for the details. After the object is created, we call the `save` method of the `FuncAnimation` class to create a movie file,[4] or the usual `show()` to play it directly on the screen.

## 6.5 More Useful Array Operations

At the start of this chapter we introduced the most essential operations needed to use arrays in computations and for plotting, but NumPy arrays can do much more. Here we introduce a few additional operations that are convenient to know about when working with arrays. First, we often need to make an array of the same size as another array. This can be done in several ways, for instance, using the `zeros` function introduced above,

```
import numpy as np
x = np.linspace(0,10,101)
```

---

[3]Technically, what happens here is that we call the *constructor* of the class `FuncAnimation` to create an object of this class. We will cover classes and constructors in detail in Chapter 7, but, for now, it is sufficient to view this as a regular function call that returns an object of type `FuncAnimation`.

[4]This call relies on external software being installed and integrated with Matplotlib, so it might not work on all platforms.

```
a = zeros(x.shape, x.dtype)
```

or by copying the x array,

```
a = x.copy()
```

or by using the convenient function zeros_like,

```
a = np.zeros_like(x)   # zeros and same size as x
```

If we write a function that takes either a list or an array as an argument, but inside the function it needs to be an array, we can ensure that it is converted by using the function asarray:

```
a = asarray(a)
```

This statement will convert a to an array if needed (e.g., if a is a list or a single number), but do nothing if a is already an array.

The *list slicing* that we briefly introduced in Chapter 3 also works for arrays, and we can extract elements from an array a using a[f:t:i]. Here, the slice f:t:i implies a set of indices (from, to, increment), exactly as for lists. We can also use any list or array of integers to index into another array:

```
>>> a = linspace(1, 8, 8)
>>> a
array([ 1.,   2.,   3.,   4.,   5.,   6.,   7.,   8.])
>>> a[[1,6,7]] = 10
>>> a
array([ 1.,  10.,   3.,   4.,   5.,   6.,  10.,  10.])
>>> a[range(2,8,3)] = -2    # same as a[2:8:3] = -2
>>> a
array([ 1.,  10.,  -2.,   4.,   5.,  -2.,  10.,  10.])
```

Finally, we can use an array of Boolean expressions to pick out elements of an array, as demonstrated in the following example:

```
>>> a < 0
[False, False, True, False, False, True, False, False]
>>> a[a < 0]              # pick out all negative elements
array([-2., -2.])

>>> a[a < 0] = a.max() # if a[i]<10, set a[i]=10
>>> a
array([ 1.,  10.,  10.,   4.,   5.,  10.,  10.,  10.])
```

These indexing methods can often be quite useful, since, for efficiency, we often want to avoid for loops over arrays elements. Many operations that are naturally implemented as for loops can be replaced by creative array slicing and indexing, with potentially substantial improvements in efficiency.

**Arrays can have any dimension.** Just as lists, arrays can have more than one index. Two-dimensional arrays are particularly relevant, since these are natural representations of, for instance, a table of numbers. For instance, to represent a set of numbers such as

$$
\begin{bmatrix}
0 & 12 & -1 & 5 \\
-1 & -1 & -1 & 0 \\
11 & 5 & 5 & -2
\end{bmatrix}
$$

it is natural to use a two-dimensional array $A_{i,j}$ with the first index for the rows and the second for the columns:

$$
A =
\begin{bmatrix}
A_{0,0} & \cdots & A_{0,n-1} \\
\vdots & \ddots & \vdots \\
A_{m-1,0} & \cdots & A_{m-1,n-1}
\end{bmatrix}
$$

In Python code, two-dimensional arrays are not much different from the one-dimensional version, except for an extra index. Making, filling, and modifying a two-dimensional array is done in much the same way, as illustrated by the following example:

```
A = zeros((3,4))    # 3x4 table of numbers
A[0,0] = -1
A[1,0] =  1
A[2,0] = 10
A[0,1] = -5
...
A[2,3] = -100

# can also write (as for nested lists)
A[2][3] = -100
```

Notice the argument to the function `zeros`, which is a tuple specifying the number of rows and columns in the two-dimensional array. We can create an array of any dimension we want by passing a tuple of the correct length. It is quite common for arrays used in numerical computations to be *sparse*, that is, to have many zeros. It is therefore often convenient to use `zeros` to create an array of the right size and then fill in the nonzero values. Alternatively, we could create a nested list and convert it to an array with the `array` function used in the beginning of the chapter.

# Chapter 7
# Dictionaries and Strings

In this chapter we will mainly focus on two data types: dictionaries and strings. Dictionaries can be considered a generalization of the list data type, where the indices are not required to be integers. We have already used strings multiple times in the previous chapters, but we will revisit them here to introduce a number of new and useful functions. Both dictionaries and strings are particularly useful for reading and processing text files, and many of our examples will be related to such applications.

## 7.1 Dictionaries

In mathematics, a mapping is a relation between objects or structures that often takes the form of a function. A mapping $f$ is a rule that assigns a unique value $f(x)$ to a given input $x$. Mappings are also widely used in computer science and can be implemented in many different ways. For instance, a Python list can be viewed as a mapping between integers (list indices) and the objects contained in a list. More general mappings can be implemented using functions and if-tests; for instance, the mapping

```
'Norway' --> 'Oslo'
'Sweden' --> 'Stockholm'
'France' --> 'Paris'
```

could be implemented in Python as a function:

```
def f(x):
    if x == 'Norway':
        return 'Oslo'
    elif x == 'Sweden':
        return 'Stockholm'
    elif x == 'France':
        return 'Paris'
```

© The Author(s) 2020
J. Sundnes, *Introduction to Scientific Programming with Python*, Simula SpringerBriefs on Computing 6,
https://doi.org/10.1007/978-3-030-50356-7_7

Such an implementation is obviously not very convenient if we have a large number of input and output values, however. An alternative implementation of the mapping would be to use two lists of equal length, where, for instance, item $n$ in list `countries` corresponds to item $n$ in list `capitals`. However, since such general mappings are useful in many contexts, Python provides a special data structure for them, called a *dictionary*. Data structures similar to a dictionary are used in many programming languages, but they often have different names. Common names are associative array, symbol table, hash map, or simply map.

A dictionary can be seen as a generalization of a list, where the indices are not required to be integers, but can be any immutable Python data type. The "indices" of a dictionary are called *keys*, and through this course we will mostly use strings as dictionary keys. The dictionary implementation of the mapping above looks like

```
d = {'Norway':'Oslo','Sweden':'Stockholm','France':'Paris'}
```

and we can look up values in the dictionary just as we would in a list, using the dictionary *key* instead of an index:

```
print(d['Norway'])
```

To extend the dictionary with new values, we can simply write

```
d['Germany'] = Berlin
```

Notice this important difference between a list and a dictionary. For a list we had to use `append()` to add new elements. A dictionary has no `append` method, and to extend it we simply introduce a new key and a corresponding value.

Dictionaries can be initialized in two different ways: one is by using the curly brackets, as in the example above. Alternatively, we can use the built-in function `dict`, which takes a number of key–value pairs as arguments and returns the corresponding dictionary. The two approaches can look like

```
mydict = {'key1': value1, 'key2': value2, ...}

temps = {'Oslo': 13, 'London': 15.4, 'Paris': 17.5}

# or
mydict = dict(key1=value1, key2=value2, ...)

temps = dict(Oslo=13, London=15.4, Paris=17.5)
```

Notice the differences in syntax, particularly the different use of quotation marks. When initializing using curly brackets, we use a colon to separate the key from its corresponding value, and the key can be any immutable Python object (e.g., strings in the example above). When using the `dict` function, we pass the key-value pairs as *keyword arguments* to the function, and the keywords are converted to keys of type string. However, in both cases, the initialization involves defining a set of key–value pairs to populate the dictionary. A dictionary is simply an unordered collection of such key–value pairs.

We are used to looping over lists to access the individual elements. We can do the same with dictionaries, with the small but important difference that looping over a dictionary means looping over its keys, and not the values. If we want to access the values we need to look them up in the dictionary using the keys. For instance, generic code to print all the values of a dictionary would appear as follows:

```
for key in dictionary:
    value = dictionary[key]
    print(value)
```

A concrete example based on the example above could look like

```
temps = {'Oslo': 13, 'London': 15.4, 'Paris': 17.5, 'Madrid': 26}
for city in temps:
    print(f'The {city} temperature is temps{city}')
```

with the following output:

```
The Paris temperature is 17.5
The Oslo temperature is 13
The London temperature is 15.4
The Madrid temperature is 26
```

As mentioned above, a dictionary is an *unordered* collection of key–value pairs, meaning that the sequence of the keys in the dictionary is arbitrary. If we want to print or otherwise process the elements in a particular order, the keys first need to be sorted, for instance, using the built-in function **sorted**:

```
for city in sorted(temps):    # alphabetic sort of keys
    value = temps[city]
    print value
```

There can be applications where sorting the keys in this manner is important, but usually the order of a dictionary is insignificant. In most applications where the order of the elements is important, a list or an array is a more convenient data type than a dictionary.

**Dictionaries and lists share many similarities.** Much of the functionality that we are familiar with for list also exists for dictionaries. We can, for instance, check if a dictionary has a particular key with the expression **key in dict**, which returns true or false:

```
>>> if 'Berlin' in temps:
...     print('Berlin:', temps['Berlin'])
... else:
...     print('No temperature data for Berlin')
...
No temperature data for Berlin
>>> 'Oslo' in temps      # standard Boolean expression
```

```
True
```

Deleting an element of a dictionary is done exactly the same way as with lists, using the operator `del`, and we can use `len` to check its length:

```
>>> del temps['Oslo']      # remove Oslo key and value
>>> temps
{'Paris': 17.5, 'London': 15.4, 'Madrid': 26.0}
>>> len(temps)             # no of key-value pairs in dict.
3
```

In some cases, it can be useful to access the keys or values of a dictionary as separate entities, and this can be accomplished with the methods `keys` and `values`, for instance `temps.keys()` and `temps.values()` for the case above. These methods will return *iterators*, which are list-like objects that can be looped over or converted to a list:

```
>>> for temp in temps.values():
>>>     print(temp)
...
17.5
15.4
26.0
>>> keys_list = list(temps.keys())
```

Just as with lists, when we assign an existing dictionary to a new variable, the dictionary is not copied. Instead, the new variable name becomes a *reference* to the same dictionary, and changing it will also change the original variable. The following code illustrates the behavior:

```
>>> t1 = temps
>>> t1['Stockholm'] = 10.0     # change t1
>>> temps                      # temps is also changed!
{'Stockholm': 10.0, 'Paris': 17.5, 'London': 15.4,
   'Madrid': 26.0}
>>> t2 = temps.copy()          # take a copy
>>> t2['Paris'] = 16
>>> t1['Paris']                # t1 was not changed
17.5
```

Here, the call to `temps.copy()` ensures that `t2` is a copy of the original dictionary, and not a reference, so changing it does not alter the original dictionary. Recall that lists behave in the same way:

```
>>> L = [1, 2, 3]
>>> M = L
>>> M[1] = 8
>>> L[1]
8
>>> M = L.copy() #for lists, M = L[:]   also works
>>> M[2] = 0
>>> L[2]
3
```

So far we have used texts (string objects) as keys, but the keys of a dictionary can be any *immutable* (constant) object. For instance, we can use integers, floats, and tuples as keys, but not lists since they are mutable objects:

```
>>> d = {1: 34, 2: 67, 3: 0}    # key is int
>>> d = {13: 'Oslo', 15.4: 'London'} # possible
>>> d = {(0,0): 4, (1,-1): 5}   # key is tuple
>>> d = {[0,0]: 4, [-1,1]: 5}   # list is mutable/changeable
...
TypeError: unhashable type: 'list'
```

Of course, the fact that these alternatives work in Python does not mean that they are recommended or very useful. It is, for instance, hard to imagine a useful application for a dictionary with a temperature as the key and a city name as the value. Strings are the most obvious and common data type for dictionary keys and will also be the most common through this book. However, there are applications where other types of keys can be useful, as we see in the following examples.

## 7.2 Example: A Dictionary for Polynomials

The information in a polynomial such as

$$p(x) = -1 + x^2 + 3x^7$$

can be represented by a dictionary with the power as the key (`int`) and the coefficient as the value (`float` or `int`):

```
p = {0: -1, 2: 1, 7: 3}
```

More generally, a polynomial written on the form

$$p(x) = \sum_{i \in I} c_i x^i,$$

for some set of integers $I$ can be represented by a dictionary with keys $i$ and values $c_i$. To evaluate a polynomial represented by such a dictionary, we need to iterate over the keys of the dictionary, extract the corresponding values, and sum up the terms. The following function takes two arguments – a dictionary `poly` and a number or array `x` – and evaluates the polynomial in `x`:

```
def eval_poly_dict(poly, x):
    sum = 0.0
    for power in poly:
        sum += poly[power]*x**power
```

```
    return sum
```

We see that the function follows our standard recipe for evaluating a sum; set a summation variable to zero and then add in all the terms using a for loop. We can write an even shorter version of the function using Python's built-in function sum:

```
def eval_poly_dict(poly, x):
    # Python's sum can add elements of an iterator
    return sum(poly[power]*x**power for power in poly)
```

Since the keys of the polynomial dictionary are integers, we can also replace the dictionary with a list, where the list index corresponds to the power of the respective term. The polynomial above, that is, $-1 + x^2 + 3x^7$ can be represented as the list

```
p = [-1, 0, 1, 0, 0, 0, 0, 3]
```

and the general polynomial $\sum_{i=0}^{N} c_i x^i$ is stored as [c0, c1, c2, ..., cN]. The function to evaluate a polynomial represented by a list is nearly identical to the function for the dictionary. The function

```
def eval_poly_list(poly, x):
    sum = 0
    for power in range(len(poly)):
        sum += poly[power]*x**power
    return sum
```

will evaluate a polynomial $\sum_{i=0}^{N} c_i x^i$ for a given $x$. An alternative and arguably more "Pythonic version" uses the convenient enumerate function:

```
def eval_poly_list_enum(poly, x):
    sum = 0
    for power, coeff in enumerate(poly):
        sum += coeff*x**power
    return sum
```

The enumerate function essentially turns a list into a list of 2-tuples, where the first element is the index of a list element and the second is the element itself. The function is quite convenient for iterating through a list when we also need access to the indices, and it is very common in Python programs.

The representations based on dictionaries and lists are very similar, but the list representation has the obvious disadvantage that we need to store all the zeros. For "sparse" high-order polynomials, this can be quite inconvenient, and the dictionary representation is obviously better. The dictionary representation can also easily handle negative powers, for instance $\frac{1}{2}x^{-3} + 2x^4$:

```
p = {-3: 0.5, 4: 2}
print eval_poly_dict(p, x=4)
```

This code will work just fine without any modifications of the eval_poly_dict function. Lists in Python cannot have negative indices (since indexing a list with a negative number implies counting indices from the end of the list), and extending the list representation to handle negative powers is not a trivial task.

## 7.3 Example: Reading File Data to a Dictionary

Say we have a file `deg2.txt`, containing temperature data for a number of cities:

```
Oslo:           21.8
London:         18.1
Berlin:         19
Paris:          23
Rome:           26
Helsinki:       17.8
```

We now want to read this file and store the information in a dictionary, with the city names as keys and the temperatures as values. The recipe is nearly identical to the one we previously used to read file data into lists: first create an empty dictionary and then fill it with values read from the file:

```
with open('deg2.txt', 'r') as infile:
    temps = {}                    # start with empty dict
    for line in infile:
        city, temp = line.split()
        city = city[:-1]          # remove last char (:)
        temps[city]  = float(temp)
```

The only real difference between this code and previous examples based on lists is the way new data are added to the dictionary. We used the `append` method to populate an empty list, but dictionaries have no such method. Instead, we add a new key–value pair with the line `temps[city] = float(temp)`. Apart from this technical difference, the recipe for populating a dictionary is exactly the same as for lists.

## 7.4 String Manipulation

We have already worked with strings (having type `str`)in previous chapters, for instance introducing the very useful `split`-method:

```
>>> s = 'This is a string'
>>> s.split()
['This', 'is', 'a', 'string']
```

String manipulation is essential for reading and interpreting the content of files, and the way we process files often depends on the file structure. For instance, we need to know the line on which the relevant information starts, how data items are separated, and how many data items are on each line. The algorithm for reading and processing the text often needs to be tailored to the file structure. Although the `split` function already considered is quite flexible, and works for most of the examples presented in this book, it might not always be the best tool. Python has a number of other ways to process strings, which could, in some cases, make the text processing easier and more efficient.

To introduce some of the basic operations on strings, we can use the following string variable as an example:

```
>>> s = 'Berlin: 18.4 C at 4 pm'
```

Such a string is really just a sequence of characters, and it behaves much like other sequence data types such as lists and tuples. For instance, we can index a string to extract individual characters;

```
>>> s[0]
'B'
>>> s[1]
'e'
>>> s[-1]
'm'
```

Slices also work in the way we are used to and can be used to extract substrings of a string:

```
>>> s
'Berlin: 18.4 C at 4 pm'
>>> s[8:]       # from index 8 to the end of the string
'18.4 C at 4 pm'
>>> s[8:12]     # index 8, 9, 10, and 11 (not 12!)
'18.4'
>>> s[8:-1]
'18.4 C at 4 p'
>>> s[8:-8]
'18.4 C'
```

Iterating over a string also works as we would expect:

```
>>> s = 'Berlin: 18.4 C at 4 pm'
>>> for s_ in s:
        print(s_, end=' ')
```

Strings have a method named `find` that searches the string for a given substring, and returns the index of its location, as follows:

```
>>> s.find('Berlin')   # where does 'Berlin' start?
0                      # at index 0
>>> s.find('pm')
20
>>> s.find('Oslo')     # not found
-1
```

Lists do not have a `find`-method, but they have a method named `index`, which is quite similar in that it searches for a given element in the list and returns its index. Strings also have a method named `index` that does almost the same thing as `find`. However, while `find` will return $-1$ if the substring does not exist in the string, `index` will end with an error message. If we want to know if a substring is part of a string and do not really care about its location, we can also use `in`:

```
>>> 'Berlin' in s:
True
>>> 'Oslo' in s:
False

>>> if 'C' in s:
...     print 'C found'
... else:
...     print 'no C'
...
C found
```

This use of `in` to check for the existence of a single element also works for lists and tuples. For strings, the method is slightly more flexible, since we can check for a substring of arbitrary length.

In many cases, we are interested not only in finding a substring, but also in finding it and replace it with something else. For this task, we have a string method named `replace`. It takes two strings as arguments, and a call such as `s.replace(s1, s2)` will replace `s1` by `s2` everywhere in the string `s`. The following examples illustrate how this method is used:

```
>>> s = 'Berlin: 18.4 C at 4 pm'
>>> s.replace(' ', '__')
'Berlin:__18.4__C__at__4__pm'
>>> s.replace('Berlin', 'Bonn')
'Bonn: 18.4 C at 4 pm'
>>> s.replace(s[:s.find(':')], 'Bonn')
'Bonn: 18.4 C at 4 pm'
```

In the final example, we combine `find` and `replace` to replace all the text before the `':'` with `'Bonn'`. First, `s.find(':')` returns the number six, which is the index where the `':'` is found; then the slice `s[:6]` is `'Berlin'`, which is replaced by `'Bonn'`. However, one important observation in this example is that these repeated calls to `s.replace` do not change `s`, but, instead, each call returns a new string where the substrings have been replaced as requested.

**Splitting and joining strings.** We have already introduced the `split` method, which is arguably the most useful method for reading and processing text files. As we recall from Chapter 5, the call `s.split(sep)` will split the string `s` into a list of substrings separated by `sep`. The `sep` argument is optional, and if it is omitted the string is split with respect to whitespace. Consider these two simple examples to recall how `split` is used:

```
>>> s = 'Berlin: 18.4 C at 4 pm'
>>> s.split(':')
['Berlin', ' 18.4 C at 4 pm']
>>> s.split()
['Berlin:', '18.4', 'C', 'at', '4', 'pm']
```

The split method has an inverse, called join, which is used to put a list of strings together with a delimiter in between:

```
>>> strings = ['Newton', 'Secant', 'Bisection']
>>> ', '.join(strings)
'Newton, Secant, Bisection'
```

Notice that we call the join method belonging to the delimiter ', ', which is a string object, and pass the list of strings as an argument. If we want to put the same list together separated by whitespace, we would simply replace ', '.join(strings) in the example above with ' '.join(strings).

Since split and join are inverse operations, using them in sequence will give back the original string, as in the following example;

```
>>> l1 = 'Oslo: 8.4 C at 5 pm'
>>> words = l1.split()
>>> l2 = ' '.join(words)
>>> l1 == l2
True
```

A common use case for the join method is to split off a known number of words on a line. Say we want to read a file in the following format, and combine the city name and the country into a single string:

```
Tromso Norway 69.6351 18.9920 52436
Molde Norway 62.7483 7.1833 18594
Oslo Norway 59.9167 10.7500 835000
Stockholm Sweden 59.3508 18.0973 1264000
Uppsala Sweden 59.8601 17.6400 133117
```

The following code will read such a file and create a nested dictionary containing the data

```
cities = {}
with open('cities.txt') as infile:
    for line in infile:
        words = line.split()
        name = ', '.join(words[:2])
        data = {'lat': float(words[2]), 'long':float(words[3])}
        data['pop'] = int(words[4])
        cities[name] = data
```

Here the line name = ', '.join(words[:2]) will create strings such as 'Tromso, Norway', which are then used as dictionary (keys). The value associated with each key is a dictionary containing the latitude and longitude data.

In most of the examples considered so far we have mostly used split for processing text files line by line, but in some cases we have a string with a great deal of text on multiple lines and we want to split it into single lines. We can do so by using the `split` method with the appropriate separator. For instance, on Linux and Mac systems, sthe line separator is `\n`;

```
>>> t = '1st line\n2nd line\n3rd line'
>>> print t
1st line
2nd line
3rd line
>>> t.split('\n')
['1st line', '2nd line', '3rd line']
```

This example works fine on Mac or Linux, but the line separator on Windows is not `\n`, but `\r\n`, and, for a platform-independent solution, it is better to use the method `splitlines()`, which works with both line separators:

```
>>> t = '1st line\n2nd line\n3rd line'      #Unix format
>>> t.splitlines()
['1st line', '2nd line', '3rd line']
>>> t = '1st line\r\n2nd line\r\n3rd line' # Windows
>>> t.splitlines()                         # cross platform!
['1st line', '2nd line', '3rd line']
```

**Strings are constant — immutable – objects.** In many of the examples above, we highlighted the similarity between strings and lists, since we are very familiar with lists from earlier chapters. However, strings are even more similar to tuples, since they are immutable objects. We could change the elements of a list in place by indexing into the list, but this does not work for strings. Trying to assign a new value to a part of a string will result in an error message:

```
>>> s[18] = 5
...
TypeError: 'str' object does not support item assignment
```

Instead, to perform such a replacement, we can build a new string manually by adding pieces of the original string or use the `replace` method introduced above:

```
>>> # build a new string by adding pieces of s:
>>> s2 = s[:18] + '5' + s[19:]
>>> s2
'Berlin: 18.4 C at 5 pm'
>>> s2 = s.replace(s[18],5)
>>> s2
'Berlin: 18.4 C at 5 pm'
```

The fact that strings are immutable, but still have a method such as `replace`, could be confusing to some. How can we replace a substring with another if strings are immutable objects? The answer is that `replace` does not really change the original string, but returns a new one. This behavior is similar to, for instance, the call `s.split()`, which will not turn `s` into a list but, instead, will leave s unchanged and *return* a list of the substrings. Similarly, a call such as s.replace(4,5) does not change s but it will return a new string that we can assign to either s or some other variable name, as we did in the example above. The call s.replace(4,5) does nothing useful on its own, unless it is combined into an assignment such as s2 = s.replace(4,5) or s = s.replace(4,5).

**Other convenient string methods in Python.** It is often convenient to strip leading or trailing whitespace from a string, and there are methods strip(), lstrip() and rstrip() to do just this:

```
>>> s = '   text with leading/trailing space   \n'
>>> s.strip()
'text with leading/trailing space'
>>> s.lstrip()   # left strip
'text with leading/trailing space   \n'
>>> s.rstrip()   # right strip
'   text with leading/trailing space'
```

We can also check whether a string contains only numbers (digits), only space, or if a string starts or ends with a given substring:

```
>>> '214'.isdigit()
True
>>> ' 214 '.isdigit()
False
>>> '2.14'.isdigit()
False

>>> '    '.isspace()    # blanks
True
>>> ' \n'.isspace()     # newline
True
>>> ' \t '.isspace()    # TAB
True
>>> ''.isspace()        # empty string
False

>>> s.startswith('Berlin')
True
>>> s.endswith('am')
False
```

Finally, we might be interested in converting between lowercase and uppercase characters:

```
>>> s.lower()
'berlin: 18.4 c at 4 pm'
>>> s.upper()
'BERLIN: 18.4 C AT 4 PM'
```

The examples shown so far are just a few of the useful string operations defined in Python. Many more exist, but all the text processing tasks considered in this book can be accomplished with the operations listed here. Nearly all the tasks we encounter in this book can be solved by using a combination of `split` and `join` in addition to string indexing and slicing.

**Example: Reading pairs of numbers (x,y) from a file.** To summarize some string operations using an example, consider the task of reading files in the following format;

```
(1.3,0)    (-1,2)     (3,-1.5)
(0,1)      (1,0)      (1,1)
(0,-0.01)  (10.5,-1)  (2.5,-2.5)
```

We want to read these coordinate pairs, convert the numbers to floats, and store them as a list of tuples. The algorithm is similar to the way we processed files earlier:

1. Read the file line by line
2. For each line, split the line into words (each number pair)
3. For each word, strip the parentheses and split the rest with respect to comma to extract the numbers

From these operations, we can observe that the `split` function is probably a good tool, as it usually is when processing text files. To strip the parentheses from the coordinate pairs, we can, for instance, use slicing. Translated into code, the example can look as follows:

```
pairs = []   # list of (n1, n2) pairs of numbers
with open('pairs.txt', 'r') as lines:
    for line in lines:
        words = line.split()
        for word in words:
            word = word[1:-1]  # strip off parentheses
            n1, n2 = word.split(',')
            n1 = float(n1);  n2 = float(n2)
            pair = (n1, n2)
            pairs.append(pair)
```

There are multiple alternative solutions for reading a file in the given format, but this one is quite simple and also relatively robust with respect to handling different numbers of pairs on each line and variable use of whitespace.

# Chapter 8
# Classes

In this chapter, we introduce classes, which is a fundamental concept in programming. Most modern programming languages support classes or similar concepts, and we have already encountered classes earlier in this book. Recall, for instance, from Chapter 2 how we can check the type of a variable with the `type` function, and the output will be of the form `<class 'int'>`, `<class 'float'>`, and so on. This simply states that the type of an object is defined in the form of a class. Every time we create, for instance, an integer variable in our program, we create an object or *instance* of the `int` class. The class defines how the objects behave and what methods they contain. We have used a large number of different methods bound to objects, such as the `append` method for list objects and `split` for strings. All such methods are part of the definition of the class to which the object belongs. So far, we have only used Python's built-in classes to create objects, but in this chapter we will write our own classes and use them to create objects tailored to our particular needs.

## 8.1 Basics of Classes

A class packs together data and functions in a single unit. As seen in previous chapters, functions that are bound to a class or an object are usually called methods, and we will stick to this notation in the present chapter. Classes have some similarity with modules, which are also collections of variables and functions that naturally belong together. However, while there can be only a single instance of a module, we can create multiple instances of a class. Different instances of the same class can contain different data, but they all behave in the same way and have the same methods. Think of a basic Python class such as `int`; we can create many integer variables in a program, and they obviously have different values (data), but we know that they all have the same general behavior and the same set of operations defined for them.

© The Author(s) 2020
J. Sundnes, *Introduction to Scientific Programming with Python*, Simula SpringerBriefs on Computing 6,
https://doi.org/10.1007/978-3-030-50356-7_8

The same goes for more complex Python classes such as lists and strings; different objects contain different data, but they all have the same methods. The classes we create in this chapter behave in exactly the same way.

**First example: A class representing a function.** To start with a familiar example, we return to the formula calculating atmospheric pressure $p$ as a function of altitude $h$. The formula we used is a simplification of a more general *barometric formula*, given by:

$$p = p_0 e^{-Mgh/RT}, \tag{8.1}$$

where $M$ is the molar mass of air, $g$ is the gravitational constant, $R$ is the gas constant, $T$ is temperature, and $p_0$ is the pressure at sea level. We obtain the simpler formula used earlier by defining the scale height as $h_0 = RT/Mg$. It could be interesting to evaluate (8.1) for different temperatures and, for each value of $T$, to create a table or plot of how the pressure varies with altitude. For each value of $T$, we need to call the function many times, with different values of $h$. How should we implement this in a convenient way? One possible solution would be to have both $h$ and $T$ as arguments:

```
from math import exp

def barometric(h, T):
    g = 9.81        #m/(s*s)
    R = 8.314       #J/(K*mol)
    M = 0.02896     #kg/mol
    p0 = 100.0      #kPa

    return p0 * exp(-M*g*h/(R*T))
```

This solution obviously works, but if we want to call the function many times for the same value of T then we still need to pass it as an argument every time it is called. However, what if the function is to be passed as an argument to another function that expects it to take a single argument only?[1] In this case, our function with two arguments will not work. A partial solution would be to include a default value for the T argument, but we would still have a problem if we want a different value of T.

Another solution would be to have $h$ as the only argument, and $T$ as a global variable:

```
T = 245.0

def barometric(h):
    g = 9.81            #m/(s*s)
```

---

[1]This situation is quite common in Python programs. Consider, for instance, the implementation of Newton's method in Chapter 4, in the functions Newton and Newton2. These functions expect two functions as arguments (f and dfdx), and both are expected to take a single argument (x). Passing in a function that requires two or more arguments will lead to an error.

```
R = 8.314          #J/(K*mol)
M = 0.02896        #kg/mol
p0 = 100.0         #kPa


return p0 * exp(-M*g*h/(R*T))
```

We now have a function that takes a single argument, but defining T as a global variable is not very convenient if we want to evaluate y(t) for different values of T. We could also set T as a local variable inside the function and define different functions barometric1(h), barometric2(h), etc., for different values of T, but this is obviously inconvenient if we want many values of T. However, we shall see that programming with classes and objects offers exactly what we need: a convenient solution to create a family of similar functions that all have their own value of T.

As mentioned above, the idea of a class is to pack together data and methods (or functions) that naturally operate on the data. We can make a class Barometric for the formula at hand, with the variables R, T,'M', g, and p0 as data, and a method value(t) for evaluating the formula. All classes should also have a method named __init__ to initialize the variables. The following code defines our function class

```
class Barometric:
    def __init__(self, T):
        self.T = T             #K
        self.g = 9.81          #m/(s*s)
        self.R = 8.314         #J/(K*mol)
        self.M = 0.02896       #kq/mol
        self.p0 = 100.0        #kPa


    def value(self, h):
        return self.p0 * exp(-self.M*self.g*h/(self.R*self.T))
```

Having defined this class, we can create *instances* of the class with specific values of the parameter T, and then we can call the method value with h as the only argument:

```
b1 = Barometric(T=245)      # create instance (object)
p1 = b1.value(2469)         # compute function value
b2 = Barometric(T=273)
p2 = b2.value(2469)
```

These code segments introduce a number of new concepts worth dissecting. First, we have a class definition that, in Python, always starts with the word class, followed by the name of the class and a colon. The following indented block of code defines the contents of the class. Just as we are used to when we implement functions, the indentation defines what belongs inside the class definition. The first contents of our class, and of most classes, is a method with the special name __init__, which is called the *constructor* of the class. This method is automatically called every time we create an instance in the class, as in the line b1 = Barometric(T=245) above. Inside the method, we

define all the constants used in the formula – self.T, self.g, and so on –
where the prefix self means that these variables become bound to the object
created. Such bound variables are called *attributes*. Finally, we define the
method value, which evaluates the formula using the predefined and object-
bound parameters self.T, self.g, self.R, self.M, and self.p0. After
we have defined the class, every time we write a line such as

```
b1 = Barometric(T=245)
```

we create a new variable (instance) b1 of type Barometric. The line looks like
a regular function call, but, since Barometric is the definition of a class and
not a function, Barometric(T=245) is instead a call to the class' constructor.
The constructor creates and returns an instance of the class with the specified
values of the parameters, and we assign this instance to the variable b. All the
__init__ functions we encounter in this book will follow exactly the same
recipe. Their purpose is to define a number of attributes for the class, and
they will typically contain one or more lines of the form self.A = A, where
A is either an argument passed to the constructor or a value defined inside
it.

As always in programming, there are different ways to achieve the same
thing, and we could have chosen a different implementation of the class above.
Since the only argument to the constructor is T, the other attributes never
change and they could have been local variables inside the value method:

```
class Barometric1:
    def __init__(self, T):
        self.T = T              #K

    def value(self, h):
        g = 9.81; R = 9.314
        M = 0.02896; p0 = 100.0
        return p0 * exp(-M*g*h/(R*self.T))
```

Notice that, inside the value method, we only use the self prefix for T, since
this is the only variable that is a class attribute. In this version of the class
the other variables are regular local variables defined inside the method. This
class does exactly the same thing as the one defined above, and one could
argue that this implementation is better, since it is shorter and simpler than
the one above. However, defining all the physical constants in one place (in
the constructor) can make the code easier to read, and the class easier to
extend with more methods. As a third possible implementation, we could
move some of the calculations from the value method to the constructor:

```
class Barometric2:
    def __init__(self, T):
        g = 9.81            #m/(s*s)
        R = 8.314           #J/(K*mol)
        M = 0.02896         #kg/mol
        self.h0 = R*T/(M*g)
        self.p0 = 100.0        #kPa
```

```
def value(self, h):
    return self.p0 * exp(-h/self.h0)
```

In this class, we use the definition of the scale height from above and compute and store this value as an attribute inside the constructor. The attribute `self.h0` is then used inside the `value` method. Notice that the constants `g`, `R`, and `M` are, in this case, local variables in the constructor, and neither these nor `T` are stored as attributes. They are only accessible inside the constructor, while `self.p0` and `self.h0` are stored and can be accessed later from within other methods.

At this point, many will be confused by the `self` variable, and the fact that, when we define the methods `__init__` and `value` they take two arguments, but, when calling them, they take only one. The explanation for this behavior is that `self` represents the object itself, and it is automatically passed as the first argument when we call a method bound to the object. When we write

```
p1 = b1.value(2469)
```

it is equivalent to the call

```
p1 = Barometric.value(b1,2469)
```

Here we explicitly call the `value` method that belongs to the `Barometric` class and pass the instance `b1` as the first argument. Inside the method, `b1` then becomes the local variable `self`, as is usual when passing arguments to a function, and we can access its attributes `T`, `g`, and so on. Exactly the same thing happens when we call `b1.value(2469)`, but now the object `b1` is automatically passed as the first argument to the method. It looks as if we are calling the method with a single argument, but in reality it gets two.

The use of the `self` variable in Python classes has been the subject of many discussions. Even experienced programmers find it confusing, and many have questioned why the language was designed this way. There are some obvious advantages to the approach, for instance, it very clearly distinguishes between instance attributes (prefixed with `self`) and local variables defined inside a method. However, if one is struggling to see the reasoning behind the `self` variable, it is sufficient to remember the following two rules: (i) `self` is always the first argument in a method definition, but is never inserted when the method is called, and (ii) to access an attribute inside a method, the attribute needs to be prefixed with `self`.

An advantage of creating a class for our barometric function is that we can now send `b1.value` as an argument to any other function that expects a function argument `f` that takes a single argument. Consider, for instance, the following small example, where the function `make_table` prints a table of the function values for any function passed to it:

```
from math import sin, exp, pi
```

```
from numpy import linspace

def make_table(f, tstop, n):
    for t in linspace(0, tstop, n):
        print(t, f(t))

def g(t):
    return sin(t)*exp(-t)

make_table(g, 2*pi, 11)              # send ordinary function

b1 = Barometric(2469)
make_table(b1.value, 2*pi, 11)    # send class method
```

Because of how f(t) is used inside the function, we need to send make_table a function that takes a single argument. Our b1.value method satisfies this requirement, but we can still use different values of T by creating multiple instances.

**More general Python classes.** Of course, Python classes have far more general applicability than just the representation of mathematical functions. A general Python class definition follows the recipe outlined in the example above, as follows:

```
class MyClass:
    def __init__(self, p1, p2,...):
        self.attr1 = p1
        self.attr2 = p2
    ...

    def method1(self, arg):
        #access attributes with self prefix
        result = self.attr1 + ...
        ...
        #create new attributes if desired
        self.attrx = arg
        ...
        return result

    def method2(self):
        ...
        print(...)
```

We can define as many methods as we want inside the class, with or without arguments. When we create an instance of the class the methods become bound to the instance, and are accessed with the prefix, for instance, m.method2() if m is an instance of MyClass. It is common to have a constructor where attributes are initialized, but this is not a requirement. Attributes can be defined whenever desired, for instance, inside a method, as in the line self.attrx = arg in the example above, or even from outside the class:

```
m = MyClass(p1,p2, ...)
m.new_attr = p3
```

The second line here creates a new attribute `new_attr` for the instance m of `MyClass`. Such addition of attributes is entirely valid, but it is rarely good programming practice since we can end up with instances of the same class having different attributes. It is a good habit to always equip a class with a constructor and to primarily define attributes inside the constructor.

## 8.2 Protected Class Attributes

For a more classical computer science example of a Python class, let us look at a class representing a bank account. Natural attributes for such a class will be the name of the owner, the account number, and the balance, and we can include methods for deposits, withdrawals, and printing information about the account. The code for defining such a class could look like this:

```python
class BankAccount:
    def __init__(self, first_name, last_name, number, balance):
        self.first_name = first_name
        self.last_name = last_name
        self.number = number
        self.balance = balance


    def deposit(self, amount):
        self.balance += amount

    def withdraw(self, amount):
        self.balance -= amount

    def print_info(self):
        first = self.first_name; last = self.last_name
        number = self.number; bal = self.balance
        s = f'{first} {last}, {number}, balance: {balance}'
        print(s)
```

Typical use of the class could be something like the following, where we create two different account instances and call the various methods for deposits, withdrawals, and printing information:

```python
>>> a1 = Account('John', 'Olsson', '19371554951', 20000)
>>> a2 = Account('Liz', 'Olsson',  '19371564761', 20000)
>>> a1.deposit(1000)
>>> a1.withdraw(4000)
>>> a2.withdraw(10500)
>>> a1.withdraw(3500)
>>> print "a1's balance:", a1.balance
a1's balance: 13500
>>> a1.print_info()
John Olsson, 19371554951, balance: 13500
>>> a2.print_info()
```

```
Liz Olsson, 19371564761, balance: 9500
```

However, there is nothing to prevent a user from changing the attributes of
the account directly:

```
>>> a1.first_name = 'Some other name'
>>> a1.balance = 100000
>>> a1.number = '19371564768'
```

Although it can be tempting to adjust a bank account balance when needed,
it is not the intended use of the class. Directly manipulating attributes in this
way will very often lead to errors in large software systems, and is considered
a bad programming style. Instead, attributes should always be changed by
calling methods, in this case, `withdraw` and `deposit`. Many programming
languages have constructions that can limit the access to attributes from
outside the class, so that any attempt to access them will lead to an error
message when compiling or running the code. Python has no technical way
to limit attribute access, but it is common to mark attributes as *protected*
by prefixing the name with an underscore (e.g., `_name`). This convention tells
other programmers that a given attribute or method is not supposed to be
accessed from outside the class, even though it is still technically possible to
do so. An account class with protected attributes can look like the following:

```python
class BankAccountP:
    def __init__(self, first_name, last_name, number, balance):
        self._first_name = first_name
        self._last_name = name
        self._number = number
        self._balance = balance

    def deposit(self, amount):
        self._balance += amount

    def withdraw(self, amount):
        self._balance -= amount

    def get_balance(self):      # NEW - read balance value
        return self._balance

    def print_info(self):
        first = self.first_name; last = self.last_name
        number = self.number; bal = self.balance
        s = f'{first} {last}, {number}, balance: {balance}'
        print(s)
```

When using this class, it will still be technically possible to access the attributes directly, as in

```python
a1 = BankAccountP('John', 'Olsson', '19371554951', 20000)
a1._number = '19371554955'
```

However, all experienced Python programmers will know that the second
line is a serious violation of good coding practice and will look for a better

way to solve the task. When using code libraries developed by others, such conventions are risky to break, since internal data structures can change, while the *interface* to the class is more static. The convention of protected variables is how programmers tell users of the class what can change and what is static. Library developers can decide to change the internal data structure of a class, but users of the class might not even notice this change if the methods to access the data remain unchanged. Since the class interface is unchanged, users who followed the convention will be fine, but users who have accessed protected attributes directly could be in for a surprise.

## 8.3 Special Methods

In the examples above, we define a constructor for each class, identified by its special name __init__(...). This name is recognized by Python, and the method is automatically called every time we create a new instance of the class. The constructor belongs to a family of methods known as *special methods*, which are all recognized by double leading and trailing underscores in the name. The term *special methods* could be a bit misleading, since the methods themselves are not really special. The special thing about them is the name, which ensures that they are automatically called in different situations, such as the __init__ function being called when class instances are created. There are many more such special methods that we can use to create object types with very useful properties.

Consider, for instance, the first example of this chapter, where the class Barometric contained the method value(h) to evaluate a mathematical function. After creating an instance named baro, we could call the method with baro.value(t). However, it would be even more convenient if we could just write baro(t) as if the instance were a regular Python function. This behavior can be obtained by simply changing the name of the value method to one of the special method names that Python automatically recognizes. The special method name for making an instance *callable* like a regular Python function is __call__:

```python
class Barometric:
    def __init__(self, T):
        self.T = T            #K
        self.g = 9.81         #m/(s*s)
        self.R = 8.314        #J/(K*mol)
        self.M = 0.02896      #kg/mol
        self.p0 = 100.0       #kPa


    def __call__(self, h):
        return self.p0 * exp(-self.M*self.g*h/(self.R*self.T))
```

Now we can call an instance of the class `Barometric` just as any other Python function

```
baro = Barometric(245)
p = baro(2346)            #same as p = baro.__call__(2346)
```

The instance `baro` now behaves and looks like a function. The method is exactly the same as the `value` method, but creating a special method by renaming it to `__call__` produces nicer syntax when the class is used.

**Special method for printing.** We are used to printing an object `a` using `print(a)`, which works fine for Python's built-in object types such as strings and lists. However, if `a` is an instance of a class we defined ourselves, we do not obtain much useful information, since Python does not know what information to show. We can solve this problem by defining a special method named `__str__` in our class. The `__str__` method must return a string object, preferably a string that provides useful information about the object, and it should not take any arguments except `self`. For the function class seen above, a suitable `__str__` method could look like the following:

```
class Barometric:
    ...
    def __call__(self, h):
        return self.p0 * exp(-self.M*self.g*h/(self.R*self.T))

    def __str__(self):
        return f'p0 * exp(-M*g*h/(R*T)); T = {self.T}'
```

If we now call `print` for an instance of the class, the function expression and the value of `T` for that instance will be printed, as follows:

```
>>> b = Barometric(245)
>>> b(2469)
70.86738432067067
>>> print(b)
p0 * exp(-M*g*h/(R*T)); T = 245
```

**Special methods for mathematical operations.** So far we have seen three special methods, namely, `__init__`, `__call__`, and `__str__`, but there are many more. We will not cover them all in this book, but a few are worth mentioning. For instance, there are special methods for arithmetic operations, such as `__add__`, `__sub__`, `__mul__`, and so forth. Defining these methods inside our class will enable us to perform operations such as `c = a+b`, where `a,b` are instances of the class. The following are relevant arithmetic operations and the corresponding special method that they will call:

```
c = a + b    # c = a.__add__(b)

c = a - b    # c = a.__sub__(b)

c = a*b      # c = a.__mul__(b)
```

```
c = a/b      #  c = a.__div__(b)

c = a**e     #  c = a.__pow__(e)
```

It is natural, in most but not all cases, for these methods to return an object of the same type as the operands. Similarly, there are special methods for comparing objects,as follows:

```
a == b       #  a.__eq__(b)

a != b       #  a.__ne__(b)

a < b        #  a.__lt__(b)

a <= b       #  a.__le__(b)

a > b        #  a.__gt__(b)

a >= b       #  a.__ge__(b)
```

These methods should be implemented to return true or false, to be consistent with the usual behavior of the comparison operators. The actual contents of the special method are in all cases entirely up to the programmer. The only special thing about the methods is their name, which ensures that they are automatically called by various operators. For instance, if you try to multiply two objects with a statement such as c = a*b, Python will look for a method named __mul__ in the instance a. If such a method exists, it will be called with the instance b as the argument, and whatever the method __mul__ returns will be the result of our multiplication operation.

**The __repr__ special method.** The last special method we will consider here is a method named __repr__, which is similar to __str__ in the sense that it should return a string with information about the object. The difference is that, while __str__ should provide human-readable information, the __repr__ string will contain all the information necessary to recreate the object. For an object a, the __repr__ method is called if we call repr(a), where repr is a built-in function. The intended function of repr is such that eval(repr(a)) == a, that is, running the string output by a.__repr__ should recreate a. To illustrate its use, let us add a __repr__ method to the class Barometric from the start of the chapter:

```
class Barometric:
    ...
    def __call__(self, h):
        return self.p0 * exp(-self.M*self.g*h/(self.R*self.T))

    def __str__(self):
        return f'p0 * exp(-M*g*h/(R*T)); T = {self.T}'

    def __repr__(self):
```

```
        """Return code for regenerating this instance."""
        return f'Barometric({self.T})'
```

Again, we can illustrate how it works in an interactive shell:

```
>>> from tmp import *
>>> b = Barometric(271)
>>> print(b)
p0 * exp(-M*g*h/(R*T)); T = 245
>>> repr(b)
'Barometric(271)'
>>> b2 = eval(repr(b))
>>> print(b2)
p0 * exp(-M*g*h/(R*T)); T = 245
```

The last two lines confirm that the `repr` method works as intended, since
running `eval(repr(b)` returns an object identical to b. Both `__repr__` and
`__str__` return strings with information about an object, the difference being
that `__str__` gives information to be read by humans, whereas the output
of `__repr__` is intended to be read by Python.

**How to know the contents of a class.** Sometimes listing the contents of a
class can be useful, particularly for debugging. Consider the following dummy
class, which does nothing useful except to define a doc string, a constructor,
and a single attribute:

```
class A:
    """A class for demo purposes."""
    def __init__(self, value):
        self.v = value
```

If we now write `dir(A)` we see that the class actually contains a great deal
more than what we put into it, since Python automatically defines certain
methods and attributes in all classes. Most of the items listed are default
versions of special methods, which do nothing useful except to give the error
message `NotImplemented` if they are called. However, if we create an instance
of A, and use `dir` on that instance, we obtain more useful information:

```
>>> a = A(2)
>>> dir(a)
['__class__', '__delattr__', '__dict__', '__dir__', '__doc__', '__eq__',
 '__format__', '__ge__', '__getattribute__', '__gt__', '__hash__',
 '__init__', '__init_subclass__', '__le__', '__lt__', '__module__',
 '__ne__', '__new__', '__reduce__', '__reduce_ex__', '__repr__',
 '__setattr__', '__sizeof__', '__str__', '__subclasshook__',
 '__weakref__', 'v']
```

We see that the list contains the same (mostly useless) default versions of
special methods, but some of the items are more meaningful. If we continue
the interactive session to examine some of the items, we obtain

```
>>> a.__doc__
'A class for demo purposes.'
```

```
>>> a.__dict__
{'v': 2}
>>> a.v
2
>>> a.__module__
'__main__'
```

The `__doc__` attribute is the doc string we defined, while `__module__` is the name of the module to which class belongs, which is simply `__main__` in this case, since we defined it in the main program. However, the most useful item is probably `__dict__`, which is a dictionary containing the names and values of all the attributes of the object `a`. Any instance holds its attributes in the `self.__dict__` dictionary, which is automatically created by Python. If we add new attributes to the instance, they are inserted into the `__dict__`:

```
>>> a = A([1,2])
>>> print a.__dict__      # all attributes
{'v': [1, 2]}
>>> a.myvar = 10          # add new attribute (!)
>>> a.__dict__
{'myvar': 10, 'v': [1, 2]}
```

When programming with classes we are not supposed to use the internal data structures such as `__dict__` explicitly, but printing it to check the values of class attributes can be very useful if something goes wrong in our code.

# 8.4 Example: Automatic Differentiation of Functions

To provide a more relevant and useful example of a `__call__` special method, consider the task of computing the derivative of an arbitrary function. Given some mathematical function in Python, say,

```
def f(x):
    return x**3
```

we want to make a class **Derivative** and write

```
dfdx = Derivative(f)
```

so that **dfdx** behaves as a function that computes the derivative of **f(x)**. When the instance **dfdx** is created, we want to call it like a regular function to evaluate the derivative of **f** in a point **x**:

```
print(dfdx(2))    # computes 3*x**2 for x=2
```

It is tricky to create such a class using analytical differentiation rules, but we can write a generic class by using numerical differentiation:

$$f'(x) \approx \frac{f(x+h) - f(x)}{h}.$$

For a small (yet moderate) $h$, say $h = 10^{-5}$, this estimate will be sufficiently accurate for most applications. The key parts of the implementation are to let the function f be an attribute of the Derivative class and then implement the numerical differentiation formula in a __call__ special method:

```
class Derivative:
    def __init__(self, f, h=1E-5):
        self.f = f
        self.h = float(h)

    def __call__(self, x):
        f, h = self.f, self.h        # make short forms
        return (f(x+h) - f(x))/h
```

The following interactive session demonstrates typical use of the class:

```
>>> from math import *
>>> df = Derivative(sin)
>>> x = pi
>>> df(x)
-1.000000082740371
>>> cos(x)  # exact
-1.0
>>> def g(t):
...         return t**3
...
>>> dg = Derivative(g)
>>> t = 1
>>> dg(t)  # compare with 3 (exact)
3.000000248221113
```

For a particularly useful application of the Derivative class, consider the solution of a nonlinear equation $f(x) = 0$. In Chapter 4 we implement Newton's method as a general method for solving nonlinear equations, but Newton's method uses the derivative $f'(x)$, which needs to be provided as an argument to the function:

```
def Newton2(f, dfdx, x0, max_it=20, tol= 1e-3):
    ...
    return x0, converged, iter
```

See Chapter 4 for a complete implementation of the function. For many functions $f(x)$, finding $f'(x)$ can require lengthy and boring derivations, and in such cases the Derivative class is quite handy:

```
>>> def f(x):
...         return 100000*(x - 0.9)**2 * (x - 1.1)**3
...
>>> dfdx = Derivative(f)
>>> xstart = 1.01
>>> Newton2(f, dfdx, xstart)
```

```
(1.093562409134085, True, 4)
```

## 8.5 Test Functions for Classes

In Chapter 4 we introduced test functions as a method to verify that our functions were implemented correctly, and the exact same approach can be used to test the implementation of classes. Inside the test function, we define parameters for which we know the expected output, and then call our class methods and compare the results with those expected. The only additional step involved when testing classes is that we will typically create one or more instances of the class inside the test function and then call their. As an example, consider a test function for the `Derivative` class of the previous section. How can we define a test case with known output for this class? Two possible methods are; (i) to compute $(f(x+h) - f(x))/h$ by hand for some $f$ and $h$, or (ii) utilize the fact that linear functions are differentiated exactly by our numerical formula, regardless of $h$. A test function based on (ii) could look like the following:

```
def test_Derivative():
    # The formula is exact for linear functions, regardless of h
    f = lambda x: a*x + b
    a = 3.5; b = 8
    dfdx = Derivative(f, h=0.5)
    diff = abs(dfdx(4.5) - a)
    assert diff < 1E-14, 'bug in class Derivative, diff=%s' % diff
```

This function follows the standard recipe for test functions: we construct a problem with a known result, create an instance of the class, call the method, and compare the result with the expected result. However, some of the details inside the test function may be worth commenting on. First, we use a lambda function to define `f(x)`. As you may recall from Chapter 4, a lambda function is simply a compact way of defining a function, with

```
f = lambda x: a*x + b
```

being equivalent to

```
def f(x):
    return a*x + b
```

The use of the lambda function inside the test function appears straightforward at first:

```
f = lambda x: a*x + b
a = 3.5; b = 8
dfdx = Derivative(f, h=0.5)
dfdx(4.5)
```

The function f is defined to taking one argument x and also using two two local variables a and b that are defined outside the function before it is called. However, looking at this code in more detail can raise questions. Calling dfdx(4.5) implies that Derivative.__call__ is called, but how can this methods know the values of a and b when it calls our f(x) function? These variables are defined inside the test function and are therefore local, whereas the class is defined in the main program. The answer is that a function defined inside another function "remembers," or has access to, *all* the local variables of the function where it is defined. Therefore, all the variables defined inside test_Derivative become part of the *namespace* of the function f, and f can access a and b in test_Derivative even when it is called from the __call__ method in class Derivative. This construction is known as a *closure* in computer science.

## 8.6 Example: A Polynomial Class

As a summarizing example of classes and special methods, we can consider the representation of polynomials introduced in Chapter 7. A polynomial can be specified by a dictionary or list representing its coefficients and powers. For example, $1 - x^2 + 2x^3$ is

$$1 + 0 \cdot x - 1 \cdot x^2 + 2 \cdot x^3$$

and the coefficients can be stored as a list [1, 0, -1, 2]. We now want to create a class for such a polynomial and equip it with functionality to evaluate and print polynomials and to add two polynomials. Intended use of the class Polynomial could look like the following:

```
>>> p1 = Polynomial([1, -1])
>>> print(p1)
1 - x
>>> p2 = Polynomial([0, 1, 0, 0, -6, -1])
>>> p3 = p1 + p2
>>> print(p3.coeff)
[1, 0, 0, 0, -6, -1]
>>> print(p3)
1 - 6*x^4 - x^5
>>> print(p3(2.0))
-127.0
>>> p4 = p1*p2
>>> p2.differentiate()
>>> print(p2)
1 - 24*x^3 - 5*x^4
```

To make all these operations possible, the class needs the following special methods:

- `__init__`, the constructor, for the line `p1 = Polynomial([1,-1])`
- `__str__`, for doing `print(p1)`
- `__call__`, to enable the call `p3(2.0)`
- `__add__`, to make `p3 = p1 + p2` work
- `__mul__`, to allow `p4 = p1*p2`

In addition, the class needs a method `differentiate` that computes the derivative of a polynomial, and changes it in-place. Starting with the most basic methods, the constructor is fairly straightforward and the call method simply follows the recipe from Chapter 7:

```
class Polynomial:
    def __init__(self, coefficients):
        self.coeff = coefficients

    def __call__(self, x):
        s = 0
        for i in range(len(self.coeff)):
            s += self.coeff[i]*x**i
        return s
```

To enable the addition of two polynomials, we need to implement the `__add__` method, which should take one argument in addition to `self`. The method should return a new Polynomial instance, since the sum of two polynomials is a polynomial, and the method needs to implement the rules of polynomial addition. Adding two polynomials means to add terms of equal order, which, in our list representation, means to loop over the `self.coeff` lists and add individual elements, as follows:

```
class Polynomial:
    ...

    def __add__(self, other):
        # return self + other

        # start with the longest list and add in the other:
        if len(self.coeff) > len(other.coeff):
            coeffsum = self.coeff[:]   # copy!
            for i in range(len(other.coeff)):
                coeffsum[i] += other.coeff[i]
        else:
            coeffsum = other.coeff[:] # copy!
            for i in range(len(self.coeff)):
                coeffsum[i] += self.coeff[i]
        return Polynomial(coeffsum)
```

The order of the sum of two polynomials is equal to the highest order of the two, so the length of the returned polynomial must be equal to the length of the longest of the two `coeff` lists. We utilize this knowledge in the code by starting with a copy of the longest list and then looping through the shortest and adding to each element.

The multiplication of two polynomials is slightly more complex than their addition, so it is worth writing down the mathematics before implementing the __mul__ method. The formula looks like

$$\left(\sum_{i=0}^{M} c_i x^i\right)\left(\sum_{j=0}^{N} d_j x^j\right) = \sum_{i=0}^{M}\sum_{j=0}^{N} c_i d_j x^{i+j},$$

which, in our list representation, means that the coefficient corresponding to the power $i+j$ is $c_i \cdot d_j$. The list r of coefficients for the resulting polynomial should have length $N+M+1$, and an element r[k] should be the sum of all products c[i]*d[j] for which $i+j = k$. The implementation of the method could look like

```
class Polynomial:
    ...
    def __mul__(self, other):
        M = len(self.coeff) - 1
        N = len(other.coeff) - 1
        coeff = [0]*(M+N+1)   # or zeros(M+N+1)
        for i in range(0, M+1):
            for j in range(0, N+1):
                coeff[i+j] += self.coeff[i]*other.coeff[j]
        return Polynomial(coeff)
```

Just as the __add__ method, __mul__ takes one argument in addition to self, and returns a new Polynomial instance.

Turning now to the differentiate method, the rule for differentiating a general polynomial is

$$\frac{d}{dx}\sum_{i=0}^{n} c_i x^i = \sum_{i=1}^{n} i c_i x^{i-1}$$

Therefore, if c is the list of coefficients, the derivative has a list of coefficients dc, where dc[i-1] = i*c[i] for i from one to the largest index in c. Note that dc will have one element less than c, since differentiating a polynomial reduces the order by one. The full implementation of the differentiate method could look like the following:

```
class Polynomial:
    ...
    def differentiate(self):       # change self
        for i in range(1, len(self.coeff)):
            self.coeff[i-1] = i*self.coeff[i]
        del self.coeff[-1]

    def derivative(self):          # return new polynomial
        dpdx = Polynomial(self.coeff[:])   # copy
        dpdx.differentiate()
        return dpdx
```

Here, the `differentiate` method will change the polynomial itself, since this is the behavior indicated by the way the function was used above. We have also added a separate function `derivative` that does not change the polynomial but, instead, returns its derivative as a new `Polynomial` object.

Finally, let us implement the `__str__` method for printing the polynomial in human-readable form. This method should return a string representation close to the way we write a polynomial in mathematics, but achieving this can be surprisingly complicated. The following implementation does a reasonably good job:

```
class Polynomial:
    ...
    def __str__(self):
        s = ''
        for i in range(0, len(self.coeff)):
            if self.coeff[i] != 0:
                s += f' + {self.coeff[i]:g}*x^{i:g}'
        # fix layout (many special cases):
        s = s.replace('+ -', '- ')
        s = s.replace(' 1*', ' ')
        s = s.replace('x^0', '1')
        s = s.replace('x^1 ', 'x ')
        if s[0:3] == ' + ':  # remove initial +
            s = s[3:]
        if s[0:3] == ' - ':  # fix spaces for initial -
            s = '-' + s[3:]
        return s
```

For all these special methods, as well as special methods in general, it is important to be aware that their contents and behavior are entirely up to the programmer. The only *special* thing about special methods is their name, which ensures that they are automatically called by certain operations. What they actually do and what they return are decided by the programmer writing the class. If we want to write an `__add__` method that returns nothing, or returns something completely different from a sum, we are free to do so. However, it is, of course, a good habit for the `__add__(self, other)` to implement something that seems like a meaningful result of `self + other`.

# Chapter 9
# Object-Oriented Programming

Upon reading the chapter title, one could wonder why object-oriented programming (OOP) is introduced only now. We have used objects since Chapter 2, and we started making our own classes and object types in Chapter 8, so what is new in Chapter 9? The answer is that the term OOP can have two different meanings. The first simply involves programming with objects and classes, which we introduced in Chapter 8, and is more commonly referred to as object-*based* programming. The second meaning of OOP is programming with *class hierarchies*, which are families of classes that inherit their methods and attributes from each other. This is the topic of the present chapter. We will learn how to collect classes in families (hierarchies) and let child classes inherit attributes and methods from parent classes.

## 9.1 Class Hierarchies and Inheritance

A class hierarchy is a family of closely related classes organized in a hierarchical manner. A key concept is *inheritance*, which means that child classes can inherit attributes and methods from parent classes. A typical strategy is to write a general class as a base class (or parent class) and then let special cases be represented as subclasses (child classes). This approach can often save much typing and code duplication. As usual, we introduce the topic by looking at some examples.

**Classes for lines and parabolas.** As a first example, let us create a class for representing and evaluating straight lines, $y = c_0 + c_1 x$. Following the concepts and ideas introduced in Chapter 8, the implementation of the class can look like

```
import numpy as np

class Line:
```

J. Sundnes, *Introduction to Scientific Programming with Python*, Simula SpringerBriefs on Computing 6,
https://doi.org/10.1007/978-3-030-50356-7_9

```
    def __init__(self, c0, c1):
        self.c0, self.c1 = c0, c1

    def __call__(self, x):
        return self.c0 + self.c1*x

    def table(self, L, R, n):
        """Return a table with n points for L <= x <= R."""
        s = ''
        for x in np.linspace(L, R, n):
            y = self(x)
            s += f'{x:12g} {y:12g}\n'
        return s
```

We see that we have equipped the class with a standard constructor, a
__call__ special method for evaluating the linear function, and a method
table for writing a table of $x$ and $y$ values. Say we now want to write a
similar class for evaluating a parabola $y = c_0 + c_1 x + c_2 x^2$. The code could
look like

```
class Parabola:
    def __init__(self, c0, c1, c2):
        self.c0, self.c1, self.c2 = c0, c1, c2

    def __call__(self, x):
        return self.c2*x**2 + self.c1*x + self.c0

    def table(self, L, R, n):
        """Return a table with n points for L <= x <= R."""
        s = ''
        for x in linspace(L, R, n):
            y = self(x)
            s += f'{x:12g} {y:12g}\n'
        return s
```

We observe that the two classes are nearly identical, differing only in the
parts that involve c2. Although we could very quickly just copy all the code
from the Line class and edit the small parts that are needed, such duplication
of code is usually a bad idea. At some point, we may need change the code,
for instance, to correct an error or improve the functionality, and having
to make the same change in multiple places often leads to time-consuming
errors. So, is there a way we can utilize the class Line code in Parabola
without resorting to copying and pasting? This is exactly what inheritance
is about.

To introduce inheritance, let us first look at the following class definition:

```
class Parabola(Line):
    pass
```

Here pass is just a Python keyword that can be used wherever Python expects to find code, but we do not want to define anything. So, at first sight,
this Parabola class seems to be empty, but notice the class definition class

Parabola(Line), which means that Parabola is a subclass of Line and *inherits* all its methods and attributes. The new Parabola class therefore has attributes c0 and c1 and three methods __init__, __call__, and table. Line is a *base class* (or parent class, superclass) , and Parabola is a *subclass* (or child class, derived class). The new Parabola class, therefore, is not as useless as it first seemed, but it is still just a copy of the Line class. To make the class represent a parabola, we need to add the missing code, that is, the code that differs between Line and Parabola. When creating such subclasses, the principle is to reuse as much as possible from the base class, only add what is needed in the subclass, and avoid duplicating code. Inspecting the two original classes above, we see that the Parabola class must add code to Line's constructor (an extra c2 attribute) and an extra term in __call__, but table can be used unaltered. The full definition of Parabola as a subclass of Line becomes the following:

```
class Parabola(Line):
    def __init__(self, c0, c1, c2):
        super().__init__(c0, c1)  # Line stores c0, c1
        self.c2 = c2

    def __call__(self, x):
        return super().__call__(x) + self.c2*x**2
```

To maximize code reuse, we allow the Parabola class to call the methods from Line, and then add the missing parts. A subclass can always access its base class bt using the built-in function super(), and this is the preferred way to call methods from the base class. We could, however, also use the class name directly, for instance Line.__init__(self,c0,c1). Generally, these two methods for invoking superclass methods look like the following:

```
SuperClassName.method(self, arg1, arg2, ...)
super().method(arg1, arg2, ...)
```

Notice the difference between the two approaches. When using the class name directly, we need to include self as the first argument, whereas this aspect is handled automatically when using super(). The use of super() is usually preferred, but in most cases the two approaches are equivalent.

To summarize this first example, the main benefits of introducing the subclass are as follows:

- Class Parabola just adds code to the already existing code in class Line, with no duplication of the code for storing c0 and c1 and computing $c_0 + c_1 x$.
- Class Parabola also has a table method; it is inherited and does not need to be written.
- __init__ and __call__ are *overridden* or *redefined* in the subclass, with no code duplication.

We can use the Parabola class and call its methods just as if they were implemented in the class directly:

```
p = Parabola(1, -2, 2)
p1 = p(2.5)
print(p1)
print(p.table(0, 1, 3))
```

**The real meaning of inheritance.** From a practical viewpoint, and for the examples in this book, the point of inheritance is to reuse methods and attributes from the base class and minimize code duplication. On a more theoretical level, inheritance should be thought of as an "is-a" relationship between the the two classes. By this we mean that if `Parabola` is a subclass of `Line`, an instance of `Parabola` is also a `Line` instance. The `Parabola` class is thought of as a special case of the `Line` class, and therefore every `Parabola` is also a `Line`, but not vice versa. We can check class type and class relations with the built-in functions `isinstance(obj, type)` and `issubclass(subclassname, superclassname)`:

```
>>> from Line_Parabola import Line, Parabola
>>> l = Line(-1, 1)
>>> isinstance(l, Line)
True
>>> isinstance(l, Parabola)
False
>>> p = Parabola(-1, 0, 10)
>>> isinstance(p, Parabola)
True
>>> isinstance(p, Line)
True
>>> issubclass(Parabola, Line)
True
>>> issubclass(Line, Parabola)
False
>>> p.__class__ == Parabola
True
>>> p.__class__.__name__    # string version of the class name
'Parabola'
```

We will not use these methods much in practical applications[1], but they are very useful for gaining a feel for class relationships when learning OOP.

Mathematically oriented readers might have noticed a logical fault in the small class hierarchy we have presented so far. We stated that a subclass is usually thought of as a special case of the base class, but a parabola is not really a special case of a straight line. It is the other way around, since a line $c_0 + c_1 x$ is a parabola $c_0 + c_1 x + c_2 x^2$ with $c_2 = 0$. Could then `Line`, then, be a subclass of `Parabola`? Certainly, and many will prefer this relation between a line and a parabola, since it follows the usual is-a relationship between a subclass and its base. The code can look like:

---

[1]If you have to use `isinstance` in your code to check what kind of object you are working with, it is usually a sign that the program is poorly designed. There are exceptions, but normally `isinstance` and `issubclass` should only be used for learning and debugging.

```
class Parabola:
    def __init__(self, c0, c1, c2):
        self.c0, self.c1, self.c2 = c0, c1, c2

    def __call__(self, x):
        return self.c2*x**2 + self.c1*x + self.c0

    def table(self, L, R, n):
        """Return a table with n points for L <= x <= R."""
        s = ''
        for x in linspace(L, R, n):
            y = self(x)
            s += '%12g %12g\n' % (x, y)
        return s

class Line(Parabola):
    def __init__(self, c0, c1):
        super().__init__(c0, c1, 0)
```

Notice that this version allows even more code reuse than the previous one, since both `__call__` and `table` can be reused without changes.

## 9.2 Example: Classes for Numerical Differentiation

Common tasks in scientific computing, such as differentiation and integration, can be carried out with a large variety of numerical methods. Many such methods are closely related, and can be easily grouped into families of methods that are very suitable for implementation in a class hierarchy. As a first example, we consider methods for numerical differentiation. The simplest formula is a one-sided finite difference:

$$f'(x) \approx \frac{f(x+h) - f(x)}{h},$$

which can be implemented in the following class:

```
class Derivative:
    def __init__(self, f, h=1E-5):
        self.f = f
        self.h = float(h)

    def __call__(self, x):
        f, h = self.f, self.h      # make short forms
        return (f(x+h) - f(x))/h
```

To use the `Derivative` class, we simply define a function `f(x)`, create an instance of the class, and call it as if it were a regular function (effectively calling the `__call__` method behind the scenes):

```
from math import exp, sin, pi

def f(x):
    return exp(-x)*sin(4*pi*x)

dfdx = Derivative(f)
print(dfdx(1.2))
```

However, numerous other formulas can be used for numerical differentiation, for instance

$$f'(x) = \frac{f(x+h) - f(x)}{h} + \mathcal{O}(h),$$

$$f'(x) = \frac{f(x) - f(x-h)}{h} + \mathcal{O}(h),$$

$$f'(x) = \frac{f(x+h) - f(x-h)}{2h} + \mathcal{O}(h^2),$$

$$f'(x) = \frac{4}{3}\frac{f(x+h) - f(x-h)}{2h} - \frac{1}{3}\frac{f(x+2h) - f(x-2h)}{4h} + \mathcal{O}(h^4),$$

$$f'(x) = \frac{3}{2}\frac{f(x+h) - f(x-h)}{2h} - \frac{3}{5}\frac{f(x+2h) - f(x-2h)}{4h} +$$

$$\frac{1}{10}\frac{f(x+3h) - f(x-3h)}{6h} + \mathcal{O}(h^6),$$

$$f'(x) = \frac{1}{h}\left(-\frac{1}{6}f(x+2h) + f(x+h) - \frac{1}{2}f(x) - \frac{1}{3}f(x-h)\right) + \mathcal{O}(h^3).$$

We can easily make a module that offers multiple formulas, as follows:

```
class Forward1:
    def __init__(self, f, h=1E-5):
        self.f, self.h = f, h

    def __call__(self, x):
        f, h = self.f, self.h
        return (f(x+h) - f(x))/h

class Central2:
    def __init__(self, f, h=1E-5):
        self.f, self.h = f, h

    def __call__(self, x):
        f, h = self.f, self.h
        return (f(x+h) - f(x-h))/(2*h)

class Central4:
    def __init__(self, f, h=1E-5):
        self.f, self.h = f, h

    def __call__(self, x):
        f, h = self.f, self.h
        return (4./3)*(f(x+h)   - f(x-h))  /(2*h) - \
```

```
(1./3)*(f(x+2*h) - f(x-2*h))/(4*h)
```

The problem with this code is, of course, that all the constructors are identical, so we duplicate a great deal of code. Although the duplication of this simple constructor might not be a big problem, it can easily lead to errors if we want to change the constructor later, and it is therefore worth avoiding. As mentioned above, a general idea of OOP is to place code common to many classes in a superclass and to have that code be inherited by the subclasses. In this case, we can make a superclass containing the constructor and let the different subclasses implement their own version of the __call__ method. The superclass will be very simple and not really useful on its own:

```
class Diff:
    def __init__(self, f, h=1E-5):
        self.f, self.h = f, h
```

The subclasses for the first-order forward formula and the second- and fourth-order central difference formulas can then look like

```
class Forward1(Diff):
    def __call__(self, x):
        f, h = self.f, self.h
        return (f(x+h) - f(x))/h

class Central2(Diff):
    def __call__(self,x):
        f, h = self.f, self.h
        return (f(x+h)-f(x-h))/(2*h)

class Central4(Diff):
    def __call__(self, x):
        f, h = self.f, self.h
        return (4./3)*(f(x+h)   - f(x-h))  /(2*h) - \
               (1./3)*(f(x+2*h) - f(x-2*h))/(4*h)
```

To use this simple class hierarchy in an example, say, we want to compute the derivative of $f(x) = \sin x$ for $x = \pi$ with the fourth-order central difference formula:

```
from math import sin, pi
mycos = Central4(sin)
mycos(pi)
```

Here, the line `mycos = Central4(sin)` creates an instance of the `Central4` class by calling the constructor inherited from the superclass, while `mycos(pi)` calls the __call__ method implemented in the subclass.

As indicated by the $O(h^n)$ terms in the formulas above, the methods have different levels of accuracy. We can empirically investigate the accuracy of the numerical differentiation formulas, using the class hierarchy created above. Using $f(x) = \sin x, x = \pi/4$ as an example, the code can look like

```
from Diff import Forward1, Central2, Central4
```

```
from math import pi, sin, cos
import numpy as np

h = [1.0/(2**i) for i in range(5)]
ref = cos(pi/4)

print(f'     h        Forward1      Central2      Central4')
for h_ in h:
    f1 = Forward1(sin,h_); c2 = Central2(sin,h_); c4 = Central4(sin,h_)
    e1 = abs(f1(pi/4)-ref)
    e2 = abs(c2(pi/4)-ref)
    e4 = abs(c4(pi/4)-ref)
    print(f'{h_:1.8f}  {e1:1.10f}  {e2:>1.10f}  {e4:>1.10f}')
```

| h | Forward1 | Central2 | Central4 |
|---|---|---|---|
| 1.00000000 | 0.4371522985 | 0.1120969417 | 0.0209220579 |
| 0.50000000 | 0.2022210836 | 0.0290966823 | 0.0014299292 |
| 0.25000000 | 0.0952716617 | 0.0073427121 | 0.0000913886 |
| 0.12500000 | 0.0459766451 | 0.0018399858 | 0.0000057438 |
| 0.06250000 | 0.0225501609 | 0.0004602661 | 0.0000003595 |

Notice that we create new instances f1, c2, and c4 for each iteration of
the loop, since we want a new value of $h$ in the formula. A more elegant
solution could be to add a new method named set_stepsize(h) or similar,
that would allow us to adjust $h$ for an existing instance. Such a method
could easily be added to the superclass and inherited by all subclasses. An
examination of the output numbers confirm that the three methods behave
as expected. For each row, we reduce $h$ by a factor of two, and the errors
are reduced by a factor of about two, four, and 16, respectively. This result
is consistent with the theoretical accuracy of the formulas, which states that
the errors should be proportional to $h$, $h^2$, and $h^4$, respectively.

## 9.3 Example: Classes for Numerical Integration

Just as numerical differentiation, numerical integration is a mainstay of com-
putational mathematics. There are numerous methods to choose from, and
they can all be written on the form

$$\int_a^b f(x)dx \approx \sum_{i=0}^{n-1} w_i f(x_i).$$

and the Based on this general formula, different methods are realized by
choosing the integration points $x_i$ and associated weights $w_i$. For instance,
the trapezoidal rule has

$$x_i = a + ih, \quad w_0 = w_{n-1} = \frac{h}{2}, \quad w_i = h \ (i \neq 0, n-1),$$

with $h = (b-a)/(n-1)$, the midpoint rule has

$$x_i = a + \frac{h}{2} + ih, \quad w_i = h,$$

with $h = (b-a)/n$, and Simpson's rule has

$$x_i = a + ih, \quad h = \frac{b-a}{n-1},$$

$$w_0 = w_{n-1} = \frac{h}{6},$$

$$w_i = \frac{h}{3} \text{ for } i \text{ even}, \quad w_i = \frac{2h}{3} \text{ for } i \text{ odd}.$$

Other methods have more complicated formulas for $w_i$ and $x_i$, and some methods choose the points randomly (e.g., Monte Carlo integration).

A numerical integration formula can be implemented as a class, with $a$, $b$, and $n$ as attributes and an `integrate` method to evaluate the formula and compute the integral. As with the family of numerical differentiation methods considered above, all such classes will be quite similar. The evaluation of $\sum_j w_j f(x_j)$ is the same, and the only difference between the methods is the definition of the points and weights. Following the ideas above, it makes sense to place all common code in a superclass, and code specific to the different methods in subclasses. Here, we can put $\sum_j w_j f(x_j)$ in a superclass (method `integrate`), and let the subclasses extend this class with code specific to a specific formula, that is, the choices of $w_i$ and $x_i$. This method-specific code can be placed inside a method, for instance, named `construct_rule`. The superclass for the numerical integration hierarchy can look like

```python
class Integrator:
    def __init__(self, a, b, n):
        self.a, self.b, self.n = a, b, n
        self.points, self.weights = self.construct_method()

    def construct_method(self):
        raise NotImplementedError('no rule in class %s' % \
                                  self.__class__.__name__)

    def integrate(self, f):
        s = 0
        for i in range(len(self.weights)):
            s += self.weights[i]*f(self.points[i])
        return s

    def vectorized_integrate(self, f):
        # f must be vectorized for this to work
        return dot(self.weights, f(self.points))
```

Notice the implementation of `construct_method`, which will raise an error if it is called, indicating that the only purpose of `Integrator` is as a superclass, and it should not be used directly. Alternatively, we could, of course, just not include the `construct_method` method in the superclass at all. However, the approach used here makes it even more obvious that the class is just a superclass and that this method needs to be implemented in subclasses.

The superclass provides a common framework for implementing the different methods, which can then be realized as subclasses. The trapezoidal and midpoint methods can be implemented as follows:

```python
class Trapezoidal(Integrator):
    def construct_method(self):
        h = (self.b - self.a)/float(self.n - 1)
        x = linspace(self.a, self.b, self.n)
        w = zeros(len(x))
        w[1:-1] += h
        w[0] = h/2;   w[-1] = h/2
        return x, w

class Midpoint(Integrator):
    def construct_method(self):
        a, b, n = self.a, self.b, self.n  # quick forms
        h = (b-a)/float(n)
        x = np.linspace(a + 0.5*h, b - 0.5*h, n)
        w = np.zeros(len(x)) + h
    return x, w
```

The more complex Simpson's rule can be added in the following subclass:

```python
class Simpson(Integrator):
    def construct_method(self):
        if self.n % 2 != 1:
            print 'n=%d must be odd, 1 is 'added % self.n
            self.n += 1
        x = np.linspace(self.a, self.b, self.n)
        h = (self.b - self.a)/float(self.n - 1)*2
        w = np.zeros(len(x))
        w[0:self.n:2] = h*1.0/3
        w[1:self.n-1:2] = h*2.0/3
        w[0] /= 2
        w[-1] /= 2
        return x, w
```

Simpson's rule is more complex because it uses different weights for odd and even points. We present all the details here for completeness, but it is not really necessary to study the details of all the formulas. The important parts here are the class design and usage of the class hierarchy.

To demonstrate how the class can be used, let us compute the integral $\int_0^2 x^2 dx$ using 101 points:

```python
def f(x):
    return x*x

simpson = Simpson(0, 2, 101)
print(simpson.integrate(f))
```

```
trapez = Trapezoidal(0,2,101)
print(trapez.integrate(f))
```

The program flow in this case might not be entirely obvious. When we construct the instance with `method = Simpson(0, 2, 101)`, the superclass constructor is invoked, but this method then calls `construct_method` in class `Simpson`. The call `method.integrate(f)` then invokes the `integrate` method inherited from the superclass. However, as users of the class, none of these details really matter to us. We use the `Simpson` class just as if all the methods were implemented directly in the class, regardless of whether they are actually inherited from another class.

# Index

© The Author(s) 2020
J. Sundnes, *Introduction to Scientific Programming with Python*, Simula SpringerBriefs on Computing 6,
https://doi.org/10.1007/978-3-030-50356-7

Printed in the United States
By Bookmasters